D1240756

Acknowledgements

There are literally hundreds of people who have contributed to this book's production and success. First and foremost I thank my wife, Tone, and my friends and family for their encouragement, support, and patience. Kudos to all the audio programs and organizations who took time to provide information for the book's data base. New Ears could not exist without their contributions. Thanks to all my colleagues at the Syracuse University College of Visual and Performing Arts for a variety of assistance. Special thanks to the folks at the SU School of Music. Thanks go out to Audio Amateur Publications, MIX Bookshelf, and the Society of Professional Audio Recording Service, especially Shirley Kaye, for their continued support with distribution and promotion. A continuing thanks to my former teachers for their contributions to audio education, including Ken Pohlmann, John Monforte, Tore Skille, Bill Porter, and John Woram. Thanks to my audio colleagues from around the world who provided information concerning their corner of the globe, especially John Binder, Chris Flam, David Moore, Brian Peters, Dave Schiffman, William Seton, Keith Seppanen, and Matt Wells.

R

New Ears
The Audio Career and Education Handbook

Compiled & Edited
by
Mark Drews

New Ear Productions
Syracuse, NY 13210-2661

Drews, Mark.
New Ear: The Audio Career and Education Handbook / Mark Drews

ISBN 0-9623502-1-4
1. Sound recording- Vocational guidance 2. Music- Vocational guidance
3. Broadcasting- Vocational guidance 4. Motion pictures- Vocational guidance

The goal of New Ears is to present information about audio engineering, music recording, and other related programs and resources. No endorsement of included programs or resources is intended or implied. The information contained within was compiled from questionnaires and additional reference materials. All data is subject to the accuracy of the original resources. If any errors are contained within or if a notable program or resource was omitted, please contact Mark Drews at New Ear Productions, 1033 Euclid Avenue, Syracuse, New York 13210-2661. All comments and suggestions are welcomed.

New Ears was produced using FileMaker Pro 2.0 and Ready, Set, Go 5.14 software on an Apple Macintosh Quadra 950 computer. Pages were produced on a Apple LaserWrite Select 310.

Library of Congress Catalog Card Number: 93-092640 ISBN: 0-9623502-1-4

Printed in the USA by Thompson-Shore 1 2 3 4 5 6 7 8 9

Preface

When I compiled the first edition of New Ears five years ago, it was a direct response to students' questions concerning educational opportunities in music recording and technology. Since then, the number of programs has exploded, with new programs emerging on a regular basis. The greatest revolution in audio education has been in traditional schools of music. Faculty and administrators are finally recognizing the importance of this type of education in their overall programs. The music recording studio is becoming accepted as a complex musical instrument, with all music students learning the basics of sound recording and performance. Ironically, music recording curricula have also become models for other educational fields attempting to blend the arts, media and technology. Students graduating from strong music recording programs with well-rounded foundations in the arts and the sciences are marketable in areas extending well beyond the recording industry.

In an effort to widen New Ears' horizons, this second edition includes a variety of new programs related to the audio community. These include audio engineering, acoustics, music recording and technology, broadcasting, music industry, multimedia, and others. Our new subtitle, The Audio Career and Education Handbook, reflects this wider view.

The audio industry is in a state of constant flux, as are many of these programs. New Ears provides a snapshot of the educational opportunities available during the period when this handbook was compiled. New programs will continue to be developed, while existing programs evolve, and some program become victims of changing technology and a competitive marketplace.

Audio fields are maturing together, with benefits being noticed in a variety of areas. Graduates from some of the earliest music recording and industry programs are now prominent members in their fields. They know the advantages of an education and are some of education's biggest supporters. More importantly, these exceptional individuals are in hiring positions, improving the industry with the talent they bring into the business.

If you are considering joining this dynamic community of ours, accept the challenge and begin to prepare yourself for this passionate industry we call audio. Turn it up!

Mark Drews
June 1993

Table of Contents

Why Audio? An Introduction to the Industry

Why audio? You may already have an answer for this curious question, but if not, this handbook is a good place to begin your search for a solution. The audio industry is comprised of many industries or perhaps more accurately, the field of audio is a subfield of many others. Any field involving sound is a part of the audio industry, and though many students of sound are attracted to high-profile careers in music recording, radio and television broadcasting, sound reinforcement, and film/video production, these areas represent a small sample of career opportunities.

This guide is designed to assist you with finding educational opportunities in audio-related fields and to act as a resource of additional audio and recording industry information. New Ears was assembled in response to the number of students that continue to be faced with the difficult task of finding a logical pathway to a career in audio. New Ears also provides useful information for other students, musicians, and working engineers interested in supplementing their knowledge of audio related areas by utilizing currently available magazines, books, seminars, short courses, trade schools, and university programs.

The audio field is challenging, demanding, and diverse. Opportunities for careers vary from music engineering to audio equipment design to product sales and support and more. Due to their diverse natures, most audio fields demand expertise in several disciplines, requiring the audio practitioner to have a command of the arts and the sciences. The individuals who successfully achieve this combination are uniquely talented. These audio professionals often find themselves doing what they love, as opposed to doing what others might view as work. This may be the answer to our original question: why audio? Audio practitioners must have an artist's passion for what they pursue.

While competition for many audio careers is intense, there is always room for excellence. Because the industry is forever changing and growing, it is constantly in need of bright, energetic, new minds to further the pursuit of creative and technical excellence in audio. The experience and education necessary to prepare yourself for an audio career is as challenging, demanding, and diverse as the industry itself. You will need to be well-educated, multi-talented, dedicated, and determined to successfully compete with your peers.

It is important to remember that your education should prepare you for the future. With the continuing technological revolution in media, the audio careers of tomorrow are just beginning to take shape. New specializations will emerge as technology shifts from analog formats to digital formats. Try to keep your eyes open for these pioneering career opportunities.

Audio Fields and Careers

While its difficult to predict the careers of the future, we can look at current opportunities in audio, and rather than cover all the fields which include sound, we will focus on audio careers involving broadcasting, engineering, music, sound reinforcement, and audio for visual media. All of these areas combine the creative elements of the arts with those of the sciences.

Broadcasting

The field of broadcasting is perhaps the largest and most powerful force in electronic media. Traditional broadcasting consisted of radio and television, but is being challenged by alternative delivery methods, such as cable and satellite. Sound is a fundamental element of all these formats. Opportunities for audio careers in broadcasting include broadcast engineering, audio production for radio and television, and location audio for electronic news gathering (ENG). Broadcast engineers typically maintain broadcast facilities. They may also design and install all production and broadcasting equipment, including transmitters. Traditionally, many broadcast engineers have received much of their training in the military or in other government organizations. With the downsizing of these organizations, fewer well-trained broadcast engineers are entering the marketplace. This is further complicated by the retirement of many veteran broadcast engineers. With the impending transition of broadcasting to new technologies, opportunities for broadcast engineers and technicians trained in the new systems will increase.

Broadcasting has several strong trade associations, including the National Association of Broadcasters (NAB), the Broadcast Education Association (BEA), the Society of Broadcast Engineers (SBE), and the Society of Motion Picture and Television Engineers (SMPTE). They all actively support education in a variety of ways including scholarships and internships. The BEA is the education branch of the NAB. The SBE is very active at the local level and often offers scholarships to students at the regional level. These organizations also host regularly scheduled conferences and exhibitions and publish journals and newsletters.

Most broadcast engineering jobs require a professional license from the FCC or SBE certification. The SBE offers courses to assist engineers with obtaining this certification. Broadcast engineer and technician programs can often be found at community colleges in the form of two-year associate degree programs, and some offer certification programs.

Engineering

We will use engineering to define technically-based careers which involve research, development, design, and manufacture of audio hardware, software, and facilities. In audio, there are two primary engineering areas: audio engineering and acoustic engineering. Audio engineering is concerned with audio hardware and software, while acoustic engineering tends to deal with sound in enclosed spaces, though both fields actively interact. Audio engineers typically are educated in electrical, computer, or mechanical engineering. Many possess advanced degrees in these areas. Acoustic engineers typically specialize in acoustics at the graduate level and may have undergraduate degrees in a variety of areas, including mechanical engineering, music, physics, or architecture. Curiously, most architecture programs include little study of acoustics. Undergraduate programs in acoustics are beginning to be established.

The main trade associations in these areas are the Audio Engineering Society (AES), the Institute of Electrical and Electronic Engineers (IEEE), and the Acoustical Society of America (ASA). The AES is the primary organization for the entire audio field. The ASA is a branch of the American Institute of Physics. Both organizations actively support education and both publish educational guides, journals, and other materials. They also sponsor annual conferences and conventions. SMPTE also supports members of the engineering research and development community. In addition to organizing the engineering fields, these organizations establish technical standards for the industry.

Students interested in these technically-oriented fields should pursue formal engineering degrees. An added benefit of a traditional engineering degree is greater flexibility in the job marketplace.

Music

The audio fields dealing with music are numerous and include the "glamor" occupations to which many people are drawn. This is easily the most competitive area in audio due to the number of people attracted to the music business. The range of occupations stretches from DJs to classical recording engineers with a variety of specializations within different music professions.

The music recording field is maturing into a true profession with support from organizations such as the National Academy of Recording Arts and Sciences (NARAS) and the Society of Professional Audio Recording Services (SPARS) in the US and the Association of Professional Recording Services (APRS) in the UK. These groups have provided forums for music recording professionals to improve the recording business. All three organizations actively support education through conferences, conventions, and special programs.

The record business is supported by organizations such as the Recording Industry Association of America (RIAA), the National Association of Record Merchandisers (NARM), and the National Association of Independent Record Distributors and Manufacturers (NAIRD). The RIAA represents the recording industry in the political arena, while NARM and NAIRD represent record retailers and distributors.

In addition to the music recording areas, the music instrument business represents a sizable field. Given the number of electronic music devices available, this area also has opportunities for engineers and designers. The National Association of Music Merchandisers (NAMM) and its educational affiliate, the National Association of Music Business Institutes (NAMBI) are active organizations within this field, along with the Music and Entertainment Industry Educators Association (MEIEA).

Many of the programs offering music recording also offer music industry courses as a component of study or as an entire program. As with many of these areas, the lines of distinction between fields are often blurred.

Sound Reinforcement

The field of sound reinforcement includes occupations related to sound system operation, design, and installation. These include live sound engineers, sound contractors, and system installers. While some may put live music engineering in this category, the largest segment of this business is related to public address and background music systems.Sound support for movie theaters and stage productions could be included in this area depending on the amount of creative involvement.

The primary trade associations for this area are the National Sound and Communications Association (NSCA) and the Custom Electronic Design and Installation Association (CEDIA). These organizations provide support through conferences and continuing education.

Most sound reinforcement training is still done through on-the-job apprenticeships due to limited formal educational opportunities.

Audio for Visual Media

Audio for visual media is a broad heading for sound occupations relating to art media, video, film, dance, and the theater. Many professionals working in these areas have backgrounds in the primary field, such as film, theater, etc., but opportunities are increasing for those with specialized audio experience. This area will continue to grow as multimedia takes root.

Art media includes art-based video, computer graphics, animation, and film. These areas can be approached from an art school background and represent experimental uses of sound with other media. Many are unaware of sound artists, but they have roots in early multimedia art that combined visual arts with electronic music.

Film production has a wealth of opportunities for audio practitioners, but as with music, the competition for jobs is intense. With the advent of improved recording and playback systems, the need for higher quality film sound has mushroomed. Film sound is another area that is maturing, offering opportunities in commercial as well as independent production.

Sound for the theater and dance share elements with sound production for art media and film. Sound designers are part of the creative team and need to have strong backgrounds in the medium and their audio specialization.

New Ears contains information about additional areas not discussed in these brief descriptions. Browse through the various programs, periodicals, trade associations, and suggested readings to further broaden your view.

Getting Started: Education and Experience

If you are currently a high school student considering a career in audio, begin planning your education by utilizing resources that are immediately available. Use the following checklist as a guide:

Protect your ears and your health. As obvious as this advice sounds, there are still many people involved in audio not giving enough thought to preserving this one vital resource. Avoid extended exposure to high or constant sound pressure levels, including those experienced at loud concerts, factories, offices, and ironically enough, recording studios. Instrumentalists in large orchestras, marching bands, and jazz/rock ensembles are also at risk of hearing loss from extended exposure while performing. Think of your ears as being equivalent to a runner's legs; if you injure one or both, your active professional career is over. Your general health is important due to the sometimes stressful nature of school and the varying physical challenges of your career. Eating right and staying strong are more vital than you may realize. Many employers judge applicants on their physical well-being, understanding that a healthy employee makes for a more dependable employee.

Study now. A strong background in computers, mathematics, music, and physics is essential for those considering a career in audio. Take advantage of college preparatory courses in these subjects offered by your high school or community college. They can save you a substantial amount of money in the future. If your school offers courses in audio recording, electronic music, and radio/television broadcasting, you can get further ahead. Consider attending a summer audio workshop or a regional audio/media conference in your area. Many conferences and conventions offer special rates for students or are free of charge. These are discussed in the next section.

Use your local library. Take advantage of the wealth of magazines, books, recordings, and videos that cover audio, music, and related topics. Libraries provide economical access to resources not easily obtainable otherwise. In addition, most libraries have systems for requesting new book purchases and magazine subscriptions. A hint for finding books on audio is to look in your library's card catalog or computer-based reference system under the headings "audio", "music", and "sound recording". For more information, please read Staying Ahead of the Game later in New Ears.

Contact professional industry associations. Several professional trade associations are directly connected with the audio field, as mentioned in the introduction. Those of primary interest are the Acoustical Society of America (ASA), the Audio Engineering Society (AES), the National Academy of Recording Arts and Sciences (NARAS), the Society of Broadcast Engineers (SBE), the Society of Motion Picture and Television Engineers (SMPTE), the Society of Professional Audio Recording Services (SPARS), and the Recording Industry Association of America (RIAA). These are the fundamental governing bodies for the industry, and they offer a variety of support for audio and recording education, such as providing internship and career counseling for students, as well as curriculum consultation for audio engineering schools. For addresses of these and other related societies and associations please refer to the New Ears Master List of Professional Trade Associations. Check your phone book for listings of local sections of professional trade organizations.

Contract working professionals in your area. It is important to accurately assess your career choice. Contact local professionals by letter or phone at their workplace. Most will be happy to discuss their occupation and explain the positive aspects of their work along with the negative ones. Make as many contacts as possible to get a variety of viewpoints. If you can, find a mentor.

Get practical experience now. Formal education is only part of the picture. Hands-on experience is critical throughout your career. This is not limited to equipment. Knowing how to deal with people is the greatest challenge of all. Practical experience can take a variety of shapes, from working at a record store to volunteering at your local public broadcasting station. Try getting involved in public access cable television production in your area. If you are a young musician, actor, writer, or filmmaker, it is never too early to begin converting practice into performance. It is far easier and cheaper to be a student artist than a starving professional.

Education Options

The number of options available for audio education continues to grow as more programs are launched. The basic school types can be defined as follows:

Seminars and Workshops- Most short programs are intended to provide basic information on specific topics in a few hours or days. They are often used to refresh fundamentals or to introduce new technologies as part of a professional's continuing education. Seminars can also provide a basic introduction to different audio fields.

Trade Schools- Trade school programs are designed to train students in specific areas of the audio and recording industries, such as recording engineering or studio maintenance. Programs tend to take several weeks or months to complete, with the more comprehensive curriculums requiring up to a year of study.

University Programs- University offerings vary from individual courses through two-year associate and minor degrees to comprehensive four-year and five-year bachelor and dual degrees. Programs originate in schools of art, business, engineering, mass communications, and music, reflecting the diversity of the audio community. Opportunities for graduate studies are also available at a growing number of schools.

Selecting A Program

Seminars and short courses can provide basic exposure to the audio field and enhance your background if you are a novice, working musician, engineer, or educator, but for the student seriously interested in an audio career, the selection of a program involves many personal decisions that go beyond simply choosing between a trade school or a university program. These decisions include how much tiime, money, and energy you are willing to spend on a formal education, but regardless of which type of education best suites your needs, the following are advised:

Get as much information as possible on the programs that interest you. This includes writing or calling the schools for additional information not included in New Ears or school brochures. A phone call to a school is an efficient way of getting specific questions answered promptly. If the program emphasizes music recording or other media, ask for promotional materials such as CD or video samples of students' work.

Talk to local audio professionals and recent graduates about their educations. Recent former students can be excellent sources of information on the inner workings of various programs and the school's social environment. Professionals can give you an idea of where new hires are getting educated.

Visit the schools of your choice. This gives you a chance to talk to faculty and current students, as well as an opportunity to tour the school and its facilities. Never enroll in a school without first visiting it; what sounds attractive on paper may be a disappointment in person.

Apply early for admission to schools that have highly selective admission requirements. This includes taking the necessary admissions tests (ACT, SAT), if required, and filing all applicable scholarship or financial aid forms. This is very important if you are competing for a spot in a highly selective program. Informing admissions' personnel that you are not in need of any financial assistance can also be beneficial in some cases.

Select the best possible program that fits your personal preferences. There are countless other factors that must be weighed when choosing a school, like costs, location, and size, but if you do your research properly, one or two programs should emerge from the pack that meet your personal requirements.

Once you're in...

Get Involved! This applies to your program and to other aspects of your school's environment. The contacts made within your program and the rest of the school's community can be as important as your formal education. Some of the greatest interdisciplinary success stories have resulted from classmates combining their talents during and after completing their educations. Your dormitory neighbor could become a film maker, a broadcast personality, a professional musician, or a banker one day, and benefit from your knowledge or vice versa. These contacts will contribute to your continuing education long after school is finished.

Demand excellence from yourself. With the cost of education skyrocketing these days, you owe it to yourself, your parents, and all those paying for your schooling to seriously apply yourself toward your educational goals. School is the beginning of your education; if you cannot apply yourself, save your money and consider your alternatives because this industry requires the ability to constantly self-educate.

Demand excellence from your school. As with the above, your school should provide an education that fulfills or exceeds its promises and one that is cost effective. Know your rights as a student and take advantage of all the services offered by your school. If something does not seem right, do not be afraid to ask questions to resolve problems.

Take the time needed to obtain a complete education. Do not feel a need to rush through school to get to your career. Proper education provides a strong, mature foundation upon which a professional career can then be built. A formal education is just the beginning of a lifetime of careers.

Keep building your background with practical experience. School offers a wealth of opportunities for practical experience. Look for work-study jobs in your area. Take advantage of summer opportunites when school activities slow. Educate yourself in areas outside your main field of study by attending master classes. These all help to build your resume and list of contacts.

Enhancing Your Program of Study

To gain a wider range of experience and broaden your horizons and resumé consider the following:

Study a foreign language. Most Americans are unprepared for dealing with any language other than English, unlike most other folks in the world. Knowing a second language will enhance your employment prospects and give you other benefits in the long run.

Spend a semester overseas. This is somewhat related to the idea of studying a foreign language, but is enhanced by actually living and studying in a different culture. Many schools offer international study programs and allow students from other schools to enroll in them. Check with your school or contact the Council on International Educational Exchange, 205 East 42nd Street, New York, NY 10017 for more information. You should also consider interning overseas.

Participate in local chapters of professional trade associations. School can isolate you from the professional community, if you're not careful. Many schools are affiliated with chapters of the major professional trade associations. If your school does not have access to a local chapter, start one. Student dues are very affordable.These organizations' regional meetings provide an informal way to mingle with the professionals from your area and establish contacts with future colleagues. Check the phone book for trade associations in your area.

Try to attend at least one national or international convention a year to further enhance your knowledge of the industry. Some of the major conferences and exhibitions are hosted by the AES, APRS, NAB, NAMM, NSCA, SMPTE, and SPARS. Check the trade magazines for conference calenders and schedules. In addition to technical equipment shows, try to attend conference related to the performing arts.

Financial Aid and Industry Assistance: Scholarships and Grants from Professional Trade Associations

Before discussing industry assistance, a few helpful hints concerning tuition and financial aid need to be mentioned. In general, community college tuition represents the best bargain in education. If you are trying to save money while working towards a bachelors degree, consider starting at a two-year community college and then transfer to a four-year university program for the final two years. State universities are reasonably priced for state residents, compared to private institutions. If your program of choice is at a state university, establishing residency in the state can save you a small fortune.

Realistically assess how much money you can afford to invest in your desired education. Be careful to take into account how your financial need can impact the cost of your education. The greater your need, the more financial aid you may be able to obtain. Though it may sound strange, this can result in a expensive school costing you less or nearly the same as a less expensive one. There are several books devoted to financial aid; a list is provided at the end of this section. Your school's financial aid office is also a good place for more information.

Because audio studies cover such a wide range of education, audio students can benefit from scholarships and grants offered by a variety of organizations. You can be rewarded for your academic, athletic, music, or dramatic ability and potential. The following is a sample of programs currently available from audio-related organizations. Addresses and phone numbers for the organizations can be found in the Master Listing of Professional Trade Associations elsewhere in this book.

The **Audio Engineering Society Educational Foundation** offers educational grants to encourage talented students to enter the profession of audio engineering and related fields. Grants for graduate studies with an emphasis on audio topics are awarded annually and can be renewed. Students are selected on the basis of demonstrated interest and achievements in audio, and on faculty recommendations. The application deadline is May 1st and awards are paid directly to the winner's school in early August. Contact the AES for more information.

The **Broadcast Education Association** manages several scholarships for study in broadcasting, radio, and law. Several are presented by the National Association of Broadcasters and the others are presented by corporate sponsors. Contact the BEA for more information.

The **Down Beat Student Music Awards** competition includes scholarship money for awards in 17 different categories, including music perofrmance, composing, arranging, and recording. Awards are given in three divisions: junior high, high school, and college. Recording awards are given for Best Engineered Live Recording and Best Engineered Studio Recording. For more information contact the Down Beat Student Music Awards, 180 W Park Avenue, Elmhurst, IL 60126.

Eastman Kodak sponsors the Eastman Scholarship program for senior and graduate level students of cinematography and film production. The scholarships are worth up to $5,000 and qualified students must be nominated by a member of their schools faculty. The program is administered by the University Film and Video Foundation. For more information contact the Foundation c/o RTVF Division, University of North Texas, PO Box 13138, Denton, TX 76203-3138. The application deadline is May 1st and awards are announced in early September.

The **Foundation for Amateur Radio** administers 47 scholarships to assist licensed radio amateurs. The awards range from $500 to $2,000 and are available to students planning to pursue a full-time course of study beyond high school. For information contact FAR Scholarships, 6903 Rhode Island Avenue, College Park, MD 20740.

The **Foundation of American Women in Television and Radio** award an annual scholarship, the Sid Guber Memorial Award, to a music performance student who shows promise and exceptional talent and has financial need. The AWRT also awards several scholarships at the local level for students pursuing careers in electronic media, radio, and television. Contact the Foundation of the AWRT for information.

The **Richard Heyser Scholarship Loan Fund** is available to selected graduate students who are registered and can demonstrate financial need to pursue their audio studies. Students may borrow up to $2,500 per year, with a maximum total of $10,000. Application forms are available from the Fund's office at 10415 Fairgrove Avenue, Tujunga, CA 91042 or the AES or Syn-Aud-Con, RR 1, Box 267, Norman, IN 47264.

The **National Association of Record Merchandisers** has two scholarships programs. The first provides 18 scholarships of $6,000 for NARM member employees or relatives. The second program, the Ernest Meyers Memorial Scholarship, was funded by a contribution from the RIAA and is designated for graduate study, targeting students entering law school.

The **National Academy of Recording Arts and Sciences** administers the NARAS Grant/Research Opportunity program. The grants, worth up to $5,000 each, are designed to provide research opportunities in the creative and technical fields of recorded music and other sound applications. They have been used to a wide range of projects. The application deadline is November 1st and grants are announced inearlye early January. Contact NARAS for more information.

The **National Foundation for the Advancement in the Arts** sponsors the Arts Recognition and Talent Search for graduating high school seniors. The program enables students to share in $3 million in college scholarships and qualify for US Presidential Scholarships. The program also assists students with apprenticeships in arts festivals and summer festivals. For more information contacts the ARTS Office, 300 NE 2nd Avenue, Miami, FL 33132 or call 305-237-3416.

The **National Sound and Communications Association** sponsors the Bud McKinney Scholarship program for students considering a career in electronics. The program allows students to attend NSCA Expositions with all their expenses paid, including access to educational sessions. Those applying must have worked for a contractor-member of the NSCA. For more information contact NSCA headquarters.

The **Society of Broadcast Engineers** maintains the Ennes Educational Foundation Scholarship program for students entering the field of Broadcast technology as well as for continuing education of SBE members. The SBE also offers scholarships at the local level through their regional chapters. Contact the SBE for more information.

For additional information on financial aid look for these books at your library or book store:

The Annual Register of Grant Support, published by R.R. Bowker

The College Blue Book: Scholarships, Fellowships, Grants, and Loans, edited by Lorraine Mathies

Directory of Financial Aids for Minorities, by Gail Ann Schlachter

Directory of Financial Aids for Women, by Gail Ann Schlachter

Financing a College Education: The Essential Guide for the 90's, by Judith Margolin

Peterson's College Money Handbook, published by Petersen's Guides, Inc.

The Scholarship Book, by Cassidy and Alves

Scholarships, Fellowships and Loans, by Norman Feingold

The Student Loan Handbook, by Lana Chandler and Michael Boggs

Winning Money for College, Alan Deutschman

The Real World

The greatest challenge facing you after your education is your entry into the marketplace. Most audio fields are extremely competitive, especially for entry-level positions. This section discusses how you can better prepare yourself for internships and career opportunities.

Internships & Summer Employment

Internships are integrated into many programs in audio, especially those which are production-oriented. Internships allow students to learn on the job, while allowing employers to assess prospective employees. Many schools put internships at the end of their curriculums with the idea that outstanding students will be offered regular employment by the businesses sponsoring the internships. This is often the case when employers find individuals who fit well into their operation. Finding good employees is a major challenge for all businesses, and the audio business in no exception.

When looking for an internship, focus on those that are handled in an organized, professional manner. Make sure you understand the expectations of your host and they understand their responsiblity as a functional component of your education. If an internship seems questionable, look elsewhere.

Because the audio commuity is relatively small, in addition to being competitive, it is vital that you positively present yourself and your abilities when interning. Be excellent at whatever task you are assigned, and demonstrate a professional and responsible attitude. Word-of-mouth concerning your performance will get or loose you your next opportunity.

The Society of Professional Audio Recording Services actively organizes internships in the recording industry for its members. SPARS also developed the National Studio Exam in the 80's to enable students to test their knowledge of recording studio technology and practice. The National Studio Exam was recently discontinued to allow SPARS to devoted more attention to their internship program.

Other professional trade associations support internships in a variety of ways, from matching students with employers to funding internship expenses. Check with the major trade associations closely aligned with your field. Begin planning your internship well in advance. Major production facilities fill internship vacancies quickly. A National Directory of Internships is available from the National Society of Internships and Education, 3509 Haworth Drive, Ste 207, Raleigh, NC 27609. For those interested in opportunities in the media arts, the National Alliance of Media Arts Centers (NAMAC) publishes a guide listing member organizations from around the country. For information, contact NAMAC, 1212 Broadway, #816, Oakland, CA 94612, phone: 510-451-2717.

You may want to consider an internship in a foreign country. The International Association for the Exchange of Students for Technical Experience (IAESTE) provides on-the-job training for students in engineering, computer science, and a variety of other fields in 50 countries around the world. Juniors, seniors, and graduate students enrolled in an accredited college or university are eligible to apply. Trainees receive a stipend to cover living expenses while training. For more information, contact the IAESTE Training Program, c/o the Association for International Practical Training, 10 Corporate Center, 10400 Little Patuxent Parkway, Columbia, MD 21044-3510, phone: 301-997-2200.

The American Institute for Foreign Study also offers internship opportunities through the Richmond College International Internship Program which combines course work and internships in some of Britain's leading companies, including some in music and media. For more information, contact AIFS, Dept INT, 102 Greenwich Avenue, CT 06830, phone: 800-727-2437.

For a wealth of information on working or studying abroad, the Council on International Educational Exchange annually publishes a guide titled Work, Study, Travel Abroad: The Whole World Handbook. It is filled with opportunities from around the globe. Contact CIEE, 205 East 42nd Street, New York, NY 10017, for more information.

Many students can benefit from internships or similar experiences placed within their educational programs, perhaps during summer breaks. Other students, especially those in more traditional engineering programs, should investigate cooperative education which allows them to alternate between going to school and working in the industry. If your education is funded by scholarships or grants, and an internship is optional, consider continuing your studies to take advantage of the academic environment.

For those students seeking summer employment in audio, check with local sound reinforcement companies or entertainment venues with seasonal work. The following companies all hire a variety of performers and technicians for summer work:

Auditions USA, Room 900, 2802 Opryland Drive, Nashville, TN 37214, phone: 800-947-8243. Auditions USA hires performers and sound/lighting technicians for the Opryland and Fiesta Texas theme parks. Opryland is in Nashville, TN, and Fiesta Texas is in San Antonio, TX. They stage a variety of live music and shows. Internship credits are available through either Belmont University or the University of Texas.

Busch Gradens, One Busch Gardens Blvd., Williamsburg, VA 23187-8785, phone: 800-253-3302. Busch Gardens Entertainment hires performers and technicians for a variety of musical productions at its theme parks, including Busch Gardens- Williamsburg, Busch Gardens- Tampa, Sea Worlds, Cyprus Gardens, and Sesame Place.

Disney Music Auditions, PO Box 10,000, Lake Buena Vista, FL 32830, phone: 800-854-8671. Disney hires musicians for it All American College Studio Orchestra and Marching Show Bands. Their summer season runs 11 to 14 weeks, May to September.

Four Star Entertainment, 1565 Rte 37 W, Suite 4, Toms River, NJ, phone: 908-914-2220. Four Star Entertainment hires performers for the Six Flags Great Adventure theme park in Jackson, NJ.

Institute of Outdoor Drama, CB #3240, NCNB Plaza, University of North Carolina, Chapel Hill, NC 27599-3240, phone: 919-962-1328. The Institute hires 300 performers, designers and technicians for their summer theater program.

Interlochen Arts Academy, PO Box 199, Interlochen, MI 49643, phone: 616-276-9221. Interlochen offers one-semester internship opportunities in recording services at their renown Arts Academy. Applicants are required to be knowledgeable of classical and jazz music and have sound reinforcement experience. Summer internships are also available.

Kings Productions, 1932 Highland Avenue, Cincinnati, OH 45219, phone: 800-544-5464. Kings Productions hires over 750 performers and technicians for the Kings Island, Kings Dominion, Carowinds, Great America, and Canada's Wonderland theme parks.

These are just a few of the summer employment opportunities worth considering. Theme parks offer the opportunity to work and make friends with a variety of people pursuing entertainment professions. These co-workers can be valuable contacts later in your career.

One additional internship opportunity deserving special attention is offered by the **Banff Center for the Arts**, Box 1020-Station 8, Banff Alberta T0L 0C0 Canada, phone: 403-726-6696. Intended for mature audio technologists, Banff's Audio Associates program allows participants to collaborate with other resident artists from a wide variety of art, music, and media fields. Associates receive remitted tuition along with a cash stipend. The Center for the Arts offers extensive audio, electronic music, video and television facilities. This is a unique example of an internship aimed at seasoned professionals. Residencies typically run for several months, depending on the participant.

The Job Search

Finding your first job can be a full-time occupation in itself, demanding persistence and patience. First, map out a plan of action. Clearly define the type of position that interests you and assemble a resumé fitting the position's requirements. Keep in mind the following:

Begin your job search before completing your education. In many fields, it is common to start looking for your first professional position the summer before your last year of school. Though this is not always possible, it can encourage you to get your plans in order and possibly help you find an internship in the process.

Most job opportunities are filled by word-of-mouth. Though some jobs are posted in classified ads in magazines or trades, the majority never make it to print. Employers need to fill positions quickly and economically, and will use their network of contacts to find the appropriate person for the job. Quite often these opportunities go to people who are in the right place, at the right time.

Utilize all the resources and contacts available to you. If your school has a job placement office, take advantage of it. If you have personal contacts in the marketplace, let them know you are looking for a position. Don't be a pest, but make your contacts aware of your availability. This is another reason why building a list of contacts is critical.

Pursue opportunities with new companies. Getting involved with a new facility allows you to contribute to the growth and development of a young business without forcing you to cope with the excess baggage sometimes common to established operations. If you have an entrepreneurial spirit, consider starting your own business or working as an independent contractor. Be a pioneer!

Allow yourself time to undertake a thorough job search. Do not rush out of school into the first position that allows you to pay your student loans. Your first professional position is a launching pad to the rest of your career. Finding that first opportunity is an investment in your future, as is your education.

Be a "people person". Businesses look for different qualities in their employees, but beyond the academic qualifications, the practical experience, and the extensive resume, employers look for individuals who are good with people. This means being a good communicator, a good listener, and a team player. Being able to cope with inflated egos and questionable attitudes is a requirement for many positions in the entertainment industry.

Continuing Education

Once you have begun to establish yourself in the industry, it will become obvious that your education must continue. With the whirlwind of change surrounding audio and music technology, you could easily spend as much time trying to keep up with the changes as you spend working. Luckily, there are structured ways to continue your education.

Read the trade magazines and journals. Though some trades appear to be extended displays of manufacturers' equipment, it is important to be aware of what is going on outside your immediate environment. Many trade magazines recognize their role as vehicles for education and include technical columns and articles aimed at refreshing your knowledge.

Attend professional trade exhibitions and conferences. All the larger trade organizations include educational sessions in their conference programs. Some, such as the AES, APRS, NAB, and SMPTE, cover a variety of topics at every conference. Others, including the SBE and SPARS, offer specialized conferences on select topics. All these organizations also publish reference materials for continuing education, and some offer video or audio recordings of their conferences for those unable to attend.

Participate in a manufacturer's workshop. Several prominant audio and video manufacturers operate their own training programs for technical operation and maintenance of their products. Some of those companies offering programs are:

Analog Devices, 666 Godwin Avenue, Midland Park, NJ 07432, phone: 617-937-1430. Aimed at design engineers, Analog Devices offers one-day seminars in Advanced Linear Design throughout the world. Though they are geared toward electronic engineers, the $20 seminars can provide anyone working with amplifiers and IC with a better understanding of their applications.

Bruel & Kjaer Instruments, 185 Forest Street, Marlborough, MA 01752, FAX: 508-485-0519. Bruel & Kjaer offer a variety of regularly scheduled courses dealing with audio topics, including acoustic noise control, fundamental measurement in electroacoustics, digital signal analysis, and sound quality analysis. These courses are offered throughout the country as one-day seminars and typically cost $125.

Texas Instruments, PO Box 172228, Denver, CO 80217-9271, phone: 800-477-8924. Texas Instruments offers $25 seminars in Advanced Linear technology targeting the design engineer. The seminar's coverage of data conversion and transmission is directly related to audio applications.

Sony Professional Audio Training Group, 6500 Congress Avenue, Boca Raton, FL 33487, phone: 407-998-6680. The Sony Professional Audio Training Group offers a variety of courses aimed at those interested in the application and service of Sony's professional audio products, including their analog and digital recorders and consoles. They also offer a seminar in CD mastering formats. Programs run from one to eight days and costs $200 to $1200. The courses are ideal for those involved in maintenance.

Sony Institute of Applied Video Technology, 2021 North Western Avenue, PO Box 29906, Hollywood, CA 90029, phone: 213-462-1987. The Sony Institute offers a variety of courses for the video professional. These include courses in production, equipment operation and troubleshooting, and post-production. College credit is available through the California State University system. Courses run two to five days and cost $600 to $1700 dollars.

Yamaha Corporation's PACE School, 6600 Orangethorpe Avenue, PO Box 6600, Buena Park, CA 90622-6600, phone: 714-522-9474. Yamaha's Program for Advanced Continuing Education (PACE) offers courses in electronic keyboard repair and sound reinforcement console operation and maintenance. Courses are two to three days in length and cost $100 to $1000. These are aimed at audio technicians involved in the installation, operation, and repair of Yamaha products.

Many other manufacturers offer courses and seminars covering their products. Some offer product training through schools listed in New Ears. Call or write companies not listed here for information regarding additional programs.

Summer Workshops

Many of the programs listed in New Ears offer summer programs aimed at the working professional. If there is a school in your area offering recording classes during the academic year, chances are they offer summer courses to help cover the cost of operating their recording facility. Among those offering regularly scheduled summer opportunities are: the Aspen Music School in Colorado, the Center for Computer Research in Music and Acoustics in California, the Hartt School of Music in Connecticut, Indiana University, New York University, the Peabody Conservatory in Maryland, the International Film and Video Workshops in Maine, the University of Iowa, the University of Northern Colorado, and the Visual Studies Institute in New York. See the Featured Schools and Program section of New Ears for more information and regularly check the audio and music trades for new programs.

Staying Ahead of the Game

As previously mentioned, the ability to self-educate is a necessary requirement for all technical occupations. To assist you with this task, the following sections list reference materials covering a variety of subjects.

Selected Audio Books, Publishers & Dealers

The number of audio-related books available has recently mushroomed, with texts covering a wide range of topics. In the course of gathering data for this handbook, schools were asked to submit detailed lists of the textbooks regularly used in their programs. The following list represents the books mentioned by a majority of the responding programs.

Advanced Digital Audio, by Ken Pohlmann, Howard W. Sams, Indianapolis, IN

Audio in Media, by Stan Alton, Wadsworth Publishing Co., Belomont, CA

Audio Post-Production in Video and Film, by Tim Amyes, Focal Press, Stoneham, MA

Audio Production Techniques for Video, by David Miles Huber, Howard W. Sams, Indianapolis, IN

Audio System Design and Installation, by Phillip Giddings, Howard W. Sams, Indianapolis, IN

Handbook for Sound Engineers: The New Audio Cyclopedia, edited by Glen Ballou, Howard W. Sams, Indianapolis, IN

Handbook of Recording Engineering, by John Eargle, Van Nostrand Reinhold, New York, NY

Introduction to Professional Recording Techniques, by Bruce Bartlett, Howard W. Sams, Indianapolis, IN

Master Handbook of Acoustics, by F. Alton Everest, TAB Books, Blue Ridge Summit, PA

Microphone Handbook, by John Eargle, ELAR Publishing Co., Commack, NY

Modern Recording Techniques, by David Miles Huber and Robert Runstein, Howard W. Sams, Indianapolis, IN

Music, Sound and Technology, by John Eargle, Van Nostrand Reinhold, New York, NY

Practical Techniques for the Recording Engineer, by Sherman Keene, SKE Publishing, Torrance, CA

Principles of Digital Audio, by Ken Pohlmann, Howard W. Sams, Indianapolis, IN

Sound Recording Handbook, by John Woram, Howard W. Sams, Indianapolis, IN

The Science of Sound, by Thomas Rossing, Addison-Wesley Publishing, Reading, MA

Sound System Engineering, by Don and Carol Davis, Howard W. Sams, Indianapolis, IN

Sound Reinforcement Handbook, by Gary Davis and Ralph Jones, Yamaha Publishing

Studio Business Book, by Jim Mandell, First House Press, Los Angeles, CA

Look for these books in your library or bookstore. If you have trouble finding them, contact the publishers or audio book dealers listed below:

A-R Editions, 315 West Gorham Street, Dept. 43, Madison, WI 53703, phone: 608-836-9000. Digital audio, computer music books

Audio Book Store, 23 Hannover Drive, Unit 7, St. Catherines, Ontario, L2W 1A3, Canada, phone: 416-641-3471. Audio books, Canadian music trades

Alan Gordon Enterprises, Inc, Publications Department, 1430 Cahuenga Blvd, Hollywood, CA 90078, phone: 213-466-3561. Film, video production books

Billboard Books, P.O. Box 2016, Lakewood, New Jersey 08701, phone: 800-344-7229. Music industry books and directories

Birns & Sawyer, 1026 N Highland Avenue, Hollywood, CA 90038, phone: 213-466-8211. Film equipment, technical books

Cinemabilia, 10 W 13th Street, New York, NY 10011, phone: 212-989-8519. Film, theater, electronic media books

Edmunds Book Store, 6658 Hollywood Blvd, Hollywood, CA 90028, phone: 213-463-3273. Motion picture, theater books

ELAR Publishing Co, 38 Pine Hill Lane, Dix Hills, NY 11746, phone: 516-586-6530. Audio books, db magazine

First Light Video Publishing, 8536 Venice Blvd, Los Angeles, CA 90034. phone: 800-777-1576. Audio and video production software, videos, CDs and laserdiscs

Focal Press, 80 Montvale Ave, Stoneham, MA 021180, phone: 617-438-8464. Audio, video books

Knowledge Industry Publications, 701 Westchester Avenue, White Plains, NY 10604, phone: 914-328-9157. Broadcast, video books

Mix Bookshelf, 6400 Hollis Street, Suite 12, Emeryville, CA 94608, phone: 800-233-9604. Audio, video, electronics, music books, videos, and software, Mix magazine

Noteworthy Music, 17 Airport Road, Nashua, NH 03063, phone: 800-648-7972. Low priced CD dealer.

Howard W. Sams & Company, P.O. Box 7092, Indianapolis, IN 46207, phone: 317-298-5511. Audio, electronics, video books

Schirmer Books, 866 Third Avenue, New York, NY, FAX: 212-319-1216. Music recording, music business books

Tab Books, Blue Ridge Summit, PA 17214, phone: 717-794-2191. Audio, acoustics, electronics books

Van Nostrand Reinhold, 135 West 50th Street, New York, NY. Audio, music technology books

Wadsworth Publishing Company, 10 David Drive, Belmont, CA 94002. Mass communications, broadcast, film books.

Mix Bookshelf in California and the Audio Book Store in Canada offer comprehensive catalogs of audio-related books. Both carry a variety of publishers and ship books internationally. Phone orders are accepted when purchases are paid for with a credit card.

Suggested Reading

In addition to the titles previously mentioned, the following books are worth considering for your library. They are indexed by topic.

Audio and Music Recording History

From Tinfoil to Stereo:The Evolution of the Phonograph, by Oliver Read and Walter Welch, Howard W. Sams, Indianapolis, IN. Tracing the development of the phonograph, this book provides insight into the relationship between the elements which shape and influence emerging audio technologies.

Hit Men, by Fredric Dannen, Times Books/Random House, New York, NY. This book chronicles the evolution of America's largest record labels. Required reading for those considering a career in the record business.

Hollywood Studio Musicians: Their Work and Careers in the Recording Industry, by Robert Faulkner, University Press of America, Lanham, MA. Originally published in 1971, this study is based on research from 1965 to 1968, providing an interesting historical perspective on the studio musician's role in the industry.

Magnetic Recording Handbook, by Marvin Camras, Van Nostrand Reinhold, New York, NY. This book is a comprehensive study of magnetic recording history, development, and application.

The Media Lab: Inventing the Future at M.I.T., by Stewart Brand, Viking Penquin Inc, New York, NY. The Media Lab investigates the cutting edge of academic research and development in converging electronic media, reflecting the development of multimedia.

Music on Demand: Composers and Careers in the Hollywood Film Industry, by Robert Faulkner, Transaction Books, New Brunswick, NJ. This is an interesting look at the realities of the film and television music composition industry, exposing the social aspects of the business.

Wireless Imagination: Sound, Radio, and the Avante-garde, by Douglas Kahn and Gregory Whitehead, MIT Press, Cambridge, MA. Through a series of new essays and recently translated documents, this book traces the history of radio and sound art.

Film Sound

On the Track: A Guide to Contemporary Film Scoring, by Fred Karlin and Rayburn Wright, Schirmer Books, New York, NY. This is most comprehensive resource available on film scoring techniques, technology and business. Includes interviews with composers, film directors and editors, recording engineers, musicians, and executives.

Live Sound

Live Sound Reinforcement, by Scott Hunter Stark, TAP, Ocean Gate, NJ. An intermediate-level approach to live sound techniques, stressing theory, technology and basic operating procedures.

Microphone Techniques

Stereo Microphone Techniques, by Bruce Bartlett, Focal Press, Stoneham, MA. Bartlett's book is a comprehensive, advanced discussion of stereo recording techniques, covering theory and procedures.

Tonmeister Technology: Recording Environments, Sound Sources and Microphone Techniques, by Michael Dickreiter, translated by Stephen Temmer, Temmer Enterprises, New York, NY. Originally written for German Broadcasting System music recording engineers, this textbook is aimed at the classical music engineer.

Music Recording and Society

Big Sounds From Small Peoples: The Music Industry in Small Countries, by Krister Malm and Roger Wallis, Constable and Co., Ltd., London. Based on the findings of the Music in Small Countries project, Big Sounds From Small Peoples details the impact of the recording business on 12 diverse countries' music and culture. An excellent study of dynamic interaction.

Noise: The Political Economy of Music, by Jacques Attali, translated by Brian Massumi, University of Minnesota Press, Minneapolis, MN. Written by an economist/advisor to the French president, Noise relates the overlap of music and politics, drawing attention to the way sound and music are controlled by the institutions of society.

Sound Recording

The Master Tape Book, by Alan Parsons, Bill Foster and Chris Hollebone, APRS and BRPG, Reading, UK. A unique guide to the creation and management of audio master tapes. Covers tape formats, studio practices, tape copying, tape machine alignment, master tape care, and the producer's role.

Sound Recording Practice, edited by John Borwick for the APRS, Oxford Press, London. A multi-authored overview of audio media, including technical concerns, equipment, recording techniques, consumer products, and allied media.

Additional Career Guides

New Ears is one of a handful of hard-to-find, highly-specialized career guides. The following guides can assist you further by keeping information up-to-date and by targeting areas of specific interest. In addition to these, MIX magazine annually publishes a listing of audio recording programs in the February issue.

The AFI Guide to College Courses in Film and Television, edited by William Horrigan and Greg Beal. Information on academic emphasis, degrees, equipment, faculty, enrollment, and admissions for over 500 college and universities. American Film Institute, 2021 N Western Avenue, Los Angeles, CA 90027.

The AES Directory of Programs, prepared by the AES Education Committee. Information on audio programs, facilities, courses, and tuition, plus a helpful information on career planning and career options. Audio Engineering Society, 60 East 42nd Street, Rm 2520, New York, NY 10165-2520.

The ASA Directory of Graduate Education in Acoustics, prepared by the ASA Committe on Education in Acoustics. Lists over 90 institution's faculty members who are involved in acoustics educations. Acoustical Society of America, 500 Sunnyside Blvd, Woodbury, NY 11797.

The BEA's Report of Broadcast and Electronic Media Programs in American and Canadian Colleges and Universities, prepared by the BEA. Lists information on over 350 college and universities offering academic programs in broadcasting and electronic media. Broadcast Education Association, 1771 N Street NW, Washington, DC 20036.

Music Business Handbook & Career Guide, by David Baskerville. A comprehensive overview of the music industry, including music business, legal issues, the record industry, music in film and broadcasting, career planning and development. Sherwood Publishing Co, PO Box 85307, Los Angeles, CA 90072

The Recording Industry: A Career Handbook, edited by Bill Forman. Part of the NARAS Grammy in the Schools project. Includes basic information about careers in the music recording industry. National Academy of Recording Arts and Sciences, 303 N Glenoaks, Ste 140, Burbank, CA 91502.

Working in Hollywood: 64 Film Professionals Talk About Moviemaking, by Alexandra Brouwer and Thomas Lee Wright. Covers occupations in movie industry from top to bottom, including audio production and film scoring. Very comprehensive. Crown Publishers, 201 East 50th Street, New York, NY 10022.

New Ears Featured Schools and Programs

Schools are listed alphabetically by name.
Blank spaces indicated unsupplied information.

American River College, Music Department

Address	*4700 College Oak Drive* *Sacramento* *95841*	*CA* *USA*	
Phone	*916-484-8420*	**FAX**	
Director	*Eric Chun*	**Admission Contact**	*Amy Rodriguez*
Program Founded		**School Type**	*Community College*
Program Offered	*Commercial Music with Recording Option*		
Degrees Offered	*Associate of Arts*		
Program Length	*2 years*		
Estimated Tuition	*None*		
Main Emphasis	*Music Recording*	**Program is**	*Semi-technical*
Accreditations	*WASC*		
Number of Studios	*2*	**Is school non-profit?**	*Yes*
Types of Studios	○ *Acoustic Research* ○ *Electronic Music* ○ *Film/Foley* ○ *Radio* ○ *Audio Research* ◉ *Music Recording* ○ *Television* ○ *Video*		
Types of Recording	◉ *Analog Multitrack* ◉ *MIDI Sequencing* ○ *Video* ○ *Digital Multitrack* ○ *DAW* ○ *Film*		
Other Resources	○ *Professional Studios* ○ *Television Stations* ○ *Radio Stations* ○ *Theater Tech Dept*		
Class Size Lecture	*Varies*	**Class Size Lab**	*Varies*
Assistance	○ *Housing* ○ *Scholarships* ○ *Internships* ◉ *Financial Aid* ○ *Work-study* ○ *Job Placement*		
Admission Policy	*Open*	**Language**	*English*
Prerequisites	◉ *High School Diploma* ○ *SAT-ACT* ○ *Music Audition*		
Industry Affiliations	◉ *AES* ○ *ASA* ○ *NACB* ○ *NARAS* ○ *SMPTE* ○ *NAMBI* ○ *APRS* ○ *NAB* ○ *NAMM* ○ *SBE* ○ *SPARS* ○ *MEIEA*		
Fulltime Faculty		**Parttime Faculty**	
Faculty Awards			
Program Awards			
Student Awards			
Research areas			

American River College, Music Department

The Music Department at the American River College offers a two-year Associate of Arts in Commercial Music, with options in Performance, Recording, and Music Business. The Recording option's curriculum includes courses in music performance and theory, songwriting, recording studio techniques, music business, studio management, and general studies. Students benefit from two fully-equipped, multitrack recording studios and a variety of performing ensembles in which to participate and/or record. These groups include orchestras, concert and symphonic bands, jazz and commercial music ensembles, and choral groups. The school also offers an Associate of Arts in Music.

American River College is a public community college and charges no tuition, though students are charged a modest enrollment fee and purchase textbooks and supplies required by their course of instruction. The school admits all who are high school graduates or the equivalent, and all others 18 years of age or older. High school students under 18 can enroll with their principal's permission. Application forms are available from the school's Administrations Office, Administration Building, 4700 College Oak Drive, Sacramento, CA 95841. Written requests must include $1.25 for return postage.

American University, Physics Department

Address	*4400 Massachusetts Avenue NW* *Washington* *DC* *20016-8085* *USA*
Phone	*202-885-2743* **FAX** *202-885-3453*
Director	*Romeo Segnan* **Admission Contact** *Romeo Segnan*
Program Founded	**School Type** *University*
Program Offered	*Audio Technology/Music Technology*
Degrees Offered	*Bachelor of Science*
Program Length	*4 years*
Estimated Tuition	*$58,000*
Main Emphasis	*Audio Engineering* **Program is** *Technical*
Accreditations	
Number of Studios	*3* **Is school non-profit?** *Yes*
Types of Studios	◉ *Acoustic Research* ◉ *Electronic Music* O *Film/Foley* O *Radio* ◉ *Audio Research* ◉ *Music Recording* O *Television* O *Video*
Types of Recording	◉ *Analog Multitrack* ◉ *MIDI Sequencing* O *Video* O *Digital Multitrack* O *DAW* O *Film*
Other Resources	O *Professional Studios* O *Television Stations* O *Radio Stations* O *Theater Tech Dept*
Class Size Lecture	*Varies* **Class Size Lab** *Varies*
Assistance	◉ *Housing* ◉ *Scholarships* ◉ *Internships* ◉ *Financial Aid* ◉ *Work-study* ◉ *Job Placement*
Admission Policy	*Selective* **Language** *English*
Prerequisites	◉ *High School Diploma* ◉ *SAT-ACT* O *Music Audition*
Industry Affiliations	◉ *AES* O *ASA* O *NACB* O *NARAS* O *SMPTE* O *NAMBI* O *APRS* O *NAB* O *NAMM* O *SBE* O *SPARS* O *MEIEA*
Fulltime Faculty	*3* **Parttime Faculty** *Vari*
Faculty Awards	
Program Awards	
Student Awards	
Research areas	*Audio Techniques, Solid-State Physics, Ultrasonic Acoustics, Acoustics*

American University, Physics Department

The American University's Bachelor of Science degree in Audio Technology concentrates on electronic sound recording and reproduction, along with analog and digital music synthesis. Offered by the Department of Physics, the four-year program draws on other departments for courses in mass media, music, theater production, business, and computer science.

The program's curriculum is balanced between a general education requirement and major requirements. Topics covered include audio technology, acoustics, sound synthesis, electronics, TV studio operations, sound studio techniques, digital interfacing, physics, music theory for non-majors, and others. A Music Technology program is offered in collaboration with the Performing Arts Department for those wishing to apply audio technologies with an emphasis on performance.

The Audio Technology program benefits from three studios: a main recording studio and two electronic music studios. The main studio is professionally-equipped with a multitrack recorder, console, effects, and a full complement of microphones. The electronic music labs have a variety of analog and digital synthesizers, samplers, MIDI controllers, and computers. The main electronic music lab is SMPTE-equipped and features a variety of multitrack recorders.

A wide range of professional opportunities are availabe in the Washington, DC area, and an extensive cooperative education program allows students to gain real world experience while earning college credit. American University graduates work in nearly every audio field, including telecommunications, motion picture sound production, radio, television, and music production.

The Department of Physics counsels freshman and transfer students. Formal admission to the program requires a minimum 2.0 grade point average and approval of the department's undergraduate advisor. No musical audition is required.

Art Institute of Dallas

Address *Two NorthPark, 8080 Park Lane*
Dallas *TX*
75231 *USA*

Phone *214-692-8086* **FAX** *214-692-6541*

Director *Terry Pope* **Admission Contact** *S. Bagley*

Program Founded *1987* **School Type** *Trade School*

Program Offered *Music/Video Business*

Degrees Offered *Associate of Applied Arts*

Program Length *18 months*

Estimated Tuition *$16,000*

Main Emphasis *Audio Engineering* **Program is** *Technical*

Accreditations *CCA, SACS*

Number of Studios - **Is school non-profit?** *No*

Types of Studios ○ *Acoustic Research* ○ *Electronic Music* ○ *Film/Foley* ○ *Radio*
○ *Audio Research* ● *Music Recording* ● *Television* ● *Video*

Types of Recording ● *Analog Multitrack* ○ *MIDI Sequencing* ● *Video*
○ *Digital Multitrack* ○ *DAW* ○ *Film*

Other Resources ○ *Professional Studios* ○ *Television Stations*
○ *Radio Stations* ○ *Theater Tech Dept*

Class Size Lecture *20* **Class Size Lab** *10-20*

Assistance ● *Housing* ● *Scholarships* ● *Internships*
● *Financial Aid* ● *Work-study* ● *Job Placement*

Admission Policy *Open* **Language** *English*

Prerequisites ● *High School Diploma* ○ *SAT-ACT* ○ *Music Audition*

Industry Affiliations ○ *AES* ○ *ASA* ○ *NACB* ○ *NARAS* ● *SMPTE* ○ *NAMBI*
○ *APRS* ○ *NAB* ○ *NAMM* ○ *SBE* ● *SPARS* ○ *MEIEA*

Fulltime Faculty *10* **Parttime Faculty** *15*

Faculty Awards

Program Awards

Student Awards

Research areas

Art Institute of Dallas

The Art Institute of Dallas offers a curriculum that includes classes in studio/technical production, marketing/sales, business management, legal issues, and general education. The 18-month program is divided into 6 eleven-week quarters. The program emphasizes audio and video production and music business and management, preparing students for entry-level positions as assistant sound engineers, camera operators, video editors, sales representatives, and production assistants

Art Institute of Philadelphia

Address *1622 Chestnut Street*
Philadelphia *PA*
19103-5198 *USA*

Phone *800-275-2474* **FAX** *215-246-3339*

Director *Daniel Levinson* **Admission Contact** *Jim Palermo*

Program Founded *1989* **School Type** *Trade School*

Program Offered *Music and Video Business*

Degrees Offered *Associates Degree in Specialized Business*

Program Length *2 years*

Estimated Tuition *$21,600*

Main Emphasis *Music Business* **Program is** *Semi-technical*

Accreditations *CCA*

Number of Studios *5* **Is school non-profit?** *No*

Types of Studios O *Acoustic Research* O *Electronic Music* O *Film/Foley* O *Radio* O *Audio Research* ◉ *Music Recording* O *Television* ◉ *Video*

Types of Recording ◉ *Analog Multitrack* ◉ *MIDI Sequencing* ◉ *Video* ◉ *Digital Multitrack* ◉ *DAW* ◉ *Film*

Other Resources O *Professional Studios* ◉ *Television Stations* ◉ *Radio Stations* ◉ *Theater Tech Dept*

Class Size Lecture *30* **Class Size Lab** *20*

Assistance ◉ *Housing* ◉ *Scholarships* ◉ *Internships* ◉ *Financial Aid* ◉ *Work-study* ◉ *Job Placement*

Admission Policy *Open* **Language** *English*

Prerequisites ◉ *High School Diploma* O *SAT-ACT* O *Music Audition*

Industry Affiliations ◉ *AES* O *ASA* O *NACB* ◉ *NARAS* ◉ *SMPTE* O *NAMBI* O *APRS* O *NAB* O *NAMM* O *SBE* O *SPARS* O *MEIEA*

Fulltime Faculty *9* **Parttime Faculty** *22*

Faculty Awards

Program Awards

Student Awards

Research areas

Art Institute of Philadelphia

The Music and Video Business program prepares students for a wide range of career opportunities and is taught by an experienced group of professionals. Students learn about audio and video recording with engineers and producers, legal issues with experts in the field, and concert promotion with promoters. They study the development of a hit record and the production of live concerts and videos.

Lectures are supplemented by lab experiences in recording studios and live performance venues. Guest speakers frequently visit to lecture and share their experiences with students. The Art Institute also assists graduates in their job search by helping to determine career goals, assisting with resumes, and establishing industry contacts for a variety of entry-level positions.

Aspen Music Festival, Stanton Audio Institute

Address *250 West 54th Street, 10th Floor East*
New York *NY*
10019-5597 *USA*

Phone *212-581-2196* **FAX**

Director *John Hill* **Admission Contact** *Lee Warren*

Program Founded **School Type** *Workshop*

Program Offered *Audio Engineering*

Degrees Offered *None*

Program Length *5 weeks*

Estimated Tuition *$1,350*

Main Emphasis *Music Recording* **Program is** *Semi-technical*

Accreditations

Number of Studios *1* **Is school non-profit?** *Yes*

Types of Studios O *Acoustic Research* O *Electronic Music* O *Film/Foley* O *Radio*
O *Audio Research* ◉ *Music Recording* O *Television* O *Video*

Types of Recording ◉ *Analog Multitrack* O *MIDI Sequencing* O *Video*
◉ *Digital Multitrack* O *DAW* O *Film*

Other Resources O *Professional Studios* O *Television Stations*
O *Radio Stations* O *Theater Tech Dept*

Class Size Lecture *10* **Class Size Lab** *10*

Assistance ◉ *Housing* ◉ *Scholarships* O *Internships*
◉ *Financial Aid* O *Work-study* O *Job Placement*

Admission Policy *Highly selective* **Language** *English*

Prerequisites O *High School Diploma* O *SAT-ACT* O *Music Audition*

Industry Affiliations ◉ *AES* O *ASA* O *NACB* O *NARAS* O *SMPTE* O *NAMBI*
O *APRS* O *NAB* O *NAMM* O *SBE* O *SPARS* O *MEIEA*

Fulltime Faculty *6* **Parttime Faculty** *Varies*

Faculty Awards

Program Awards

Student Awards

Research areas

Aspen Music Festival, Stanton Audio Institute

The Aspen Music Festival and School is the site of the Edgar Stanton Audio Recording Institute each summer. The program is an intensive course in audio engineering offered during the first half-session of the Aspen Music School. Lectures and lab sessions encompass a broad range of topics related to music recording and reinforcement, with a particular emphasis on classical, live-mixing techniques.

Students gain hands-on experience in an audio control room equipped with state-of-the-art donated by major audio manufacturers. The Festival program offers a wide range of performing ensembles including symphonic orchestras, chamber ensembles, opera/choral groups, new music/computer music ensembles, big band, and jazz ensembles.

Lecture/demonstrations are conducted by ESARI faculty and prominent guest lecturers from the professional audio community. Topics may include: fundamentals of audio, electroacoustics, psychoacoustics, the recording chain, mixers and consoles, analog tape and digital recording, microphone theory, stereo mic techniques and perspectives, signal processing theory and applications, multi-microphone and live-mixing techniques, SMPTE and synchronization.

Enrollment is strictly limited to ten students to maintain a low student-to-faculty ratio. All applicants must submit a cover letter and resume along with the application for admission. The Institute does not require any specific prerequisites or training, though preference is given to applicants with some musical and/or technical background.

The Aspen Music Festival's address in Colorado is Music Associates of Aspen, PO Box AA, Aspen, CO 81612-7428.

Audio Recording Technology Institute

Address *440 Wheeler Road*
Hauppauge *NY*
11788 *USA*

Phone *516-582-8999* **FAX** *516-582-8213*

Director *James L. Bernard* **Admission Contact** *Peg Lorello*

Program Founded *1973* **School Type** *Trade School*

Program Offered *Theory and Practice of Audio Recording*

Degrees Offered *None*

Program Length *30 weeks*

Estimated Tuition *$1875-$2475*

Main Emphasis *Audio Engineering* **Program is** *Technical*

Accreditations *NY Tech University*

Number of Studios *2* **Is school non-profit?** *No*

Types of Studios
○ *Acoustic Research* ○ *Electronic Music* ○ *Film/Foley* ○ *Radio*
○ *Audio Research* ◉ *Music Recording* ○ *Television* ○ *Video*

Types of Recording
◉ *Analog Multitrack* ○ *MIDI Sequencing* ○ *Video*
○ *Digital Multitrack* ○ *DAW* ○ *Film*

Other Resources
◉ *Professional Studios* ○ *Television Stations*
○ *Radio Stations* ○ *Theater Tech Dept*

Class Size Lecture *10* **Class Size Lab** *5*

Assistance
○ *Housing* ○ *Scholarships* ○ *Internships*
◉ *Financial Aid* ○ *Work-study* ◉ *Job Placement*

Admission Policy *Highly selective* **Language** *English*

Prerequisites ◉ *High School Diploma* ○ *SAT-ACT* ○ *Music Audition*

Industry Affiliations
◉ *AES* ○ *ASA* ○ *NACB* ○ *NARAS* ○ *SMPTE* ○ *NAMBI*
○ *APRS* ○ *NAB* ○ *NAMM* ○ *SBE* ○ *SPARS* ○ *MEIEA*

Fulltime Faculty *3-4* **Parttime Faculty** *1-2*

Faculty Awards

Program Awards

Student Awards

Research areas

Audio Recording Technology Institute

The Audio Recording Technology Institute offers a comprehensive, hands-on course in audio recording theory and practice. The program is divided into the following courses: Basic 101, Advanced Audio A201, Recording Workshop A301, and MIDI Technology. Each course is 10 weeks, 3 hours per week. They provide extensive hands-on experience with studio equipment. The program maintains very limited enrollment per class and offers affordable tuition. Financing is available. Graduates of the program may use all ARTI studios for their own private sessions as member of the program's Recording Engineers Association. The Association offers graduates further professional studio experience to prepare them for employment in the industry.

ARTI also operates programs in California and Florida. The California location is at 1325 Red Gun Street, Anaheim, CA 92806, phone: 714-666-2784. The director is Tom Bernard. The Florida location is at 2307 Mt Vernon Street, Orlando, FL 32803, phone: 407-894-5400. The program's director is Jim Bernard.

Australian Audio College

Address *763 High Street, Preston, Victoria, 3042, Australia*

Phone *61-3-478-2153* **FAX** *61-3-470-1183*

Director *Rod James-Hume* **Admission Contact** *Robert Iurato*

Program Founded *1987* **School Type** *Trade School*

Program Offered *Audio Engineering*

Degrees Offered *Diploma*

Program Length *1 year*

Estimated Tuition *$2,000*

Main Emphasis *Audio Engineering* **Program is** *Technical*

Accreditations *Australian Music Exam Board*

Number of Studios *3* **Is school non-profit?** *No*

Types of Studios
- ○ Acoustic Research
- ◉ Electronic Music
- ◉ Film/Foley
- ○ Radio
- ○ Audio Research
- ○ Music Recording
- ○ Television
- ○ Video

Types of Recording
- ◉ Analog Multitrack
- ◉ MIDI Sequencing
- ◉ Video
- ○ Digital Multitrack
- ◉ DAW
- ○ Film

Other Resources
- ◉ Professional Studios
- ○ Television Stations
- ○ Radio Stations
- ○ Theater Tech Dept

Class Size Lecture *15* **Class Size Lab** *1*

Assistance
- ○ Housing
- ○ Scholarships
- ◉ Internships
- ◉ Financial Aid
- ○ Work-study
- ◉ Job Placement

Admission Policy *Selective* **Language** *English*

Prerequisites
- ○ High School Diploma
- ○ SAT-ACT
- ○ Music Audition

Industry Affiliations
- ◉ AES
- ○ ASA
- ○ NACB
- ○ NARAS
- ○ SMPTE
- ○ NAMBI
- ○ APRS
- ○ NAB
- ○ NAMM
- ○ SBE
- ○ SPARS
- ○ MEIEA

Fulltime Faculty *2* **Parttime Faculty** *2*

Faculty Awards

Program Awards

Student Awards

Research areas

Australian Audio College

The Australian Audio College admits two classes per year for its Diploma of Audio Engineering Course, the first commencing in February, and the second commencing in August of each year. The Diploma course is run over a period of 50 weeks on a part-time basis. Each student is required to attend one lecture per week for 2 1/2 hours. In addition to lectures, students are required to attend practical sessions. These are booked at times convenient for the students. Students are able to obtain as much studio time as desired, with at least 300 hours being recommended for the 50-week period. Upon successful completion of the course, the school assists students with locating suitable employment.

The Australian Audio College is registered with both the Australian Music Examination Board and the Audio Engineering Society. To date, 90% of their graduates are employed within the audio industry, and an even greater number of non-graduates are working within the industry in various positions. Melbourne's largest audio production house, Metropolis Audio, has employed several of the school's students. Others employing graduates from the Australian Audio College include Platinum Studios, Music and Special Effects Lab, Morris Studios, and the Royal Melbourne Institute of Technology.

The school places an emphasis on individual practical work in their three in-house recording studios. The school's lecturers are fully-qualified with international experience, and they try to make every lesson enjoyable. Classes are held in a relaxed environment, and numbers are kept to a minimum. If students ever need to repeat a class, they are welcomed to do so, at no further cost. The school also offers a concession rate for either full-time students or those unemployed; the course fees effectively become $1,750.

The philosophy of the College has always been simple- to make every student a successful audio engineer. They want students to not only become excellent engineers, but also employable ones. They understand their success only comes with their students' success.

Australian Film Television & Radio School

Address	*Balaclava & Epping Roads, PO Box 126* *North Ryde* *New South Wales* *2113* *Australia*		
Phone	*61-2-805-6611*	**FAX**	*61-2-887-1030*
Director	*John O'Hara*	**Admission Contact**	*Sharon Bell*
Program Founded	*1973*	**School Type**	*University*
Program Offered	*Film, Television, & Radio*		
Degrees Offered	*Bachelor of Arts in Film & Television*		
Program Length	*3 years*		
Estimated Tuition	*Inquire*		
Main Emphasis	*Film/Video Production*	**Program is**	*Technical*
Accreditations	*Minister of the Arts*		
Number of Studios	*8*	**Is school non-profit?**	*Yes*
Types of Studios	○ *Acoustic Research* ◉ *Electronic Music* ◉ *Film/Foley* ◉ *Radio* ○ *Audio Research* ◉ *Music Recording* ◉ *Television* ◉ *Video*		
Types of Recording	◉ *Analog Multitrack* ◉ *MIDI Sequencing* ◉ *Video* ○ *Digital Multitrack* ○ *DAW* ◉ *Film*		
Other Resources	◉ *Professional Studios* ◉ *Television Stations* ◉ *Radio Stations* ○ *Theater Tech Dept*		
Class Size Lecture	*Varies*	**Class Size Lab**	*Varies*
Assistance	○ *Housing* ○ *Scholarships* ○ *Internships* ◉ *Financial Aid* ○ *Work-study* ○ *Job Placement*		
Admission Policy	*Highly selective*	**Language**	*English*
Prerequisites	◉ *High School Diploma* ○ *SAT-ACT* ○ *Music Audition* *Creative Aptitude Test, Experience*		
Industry Affiliations	◉ *AES* ○ *ASA* ○ *NACB* ○ *NARAS* ◉ *SMPTE* ○ *NAMBI* ○ *APRS* ○ *NAB* ○ *NAMM* ○ *SBE* ○ *SPARS* ○ *MEIEA*		
Fulltime Faculty	*90*	**Parttime Faculty**	*2,000*
Faculty Awards			
Program Awards			
Student Awards			
Research areas			

Australian Film Television & Radio School

The Australian Film Television & Radio School is the national center for professional training in film, television and radio production. The school trains producers, directors, scriptwriters, directors of photography, camera operators, sound designers, production designers, editors and radio broadcasters. AFTRS provides fulltime training as well as a large number of short and part-time courses. They also offer postgraduate programs, commercial radio workshops, and two special programs- the Industry Training Fund for Women and the Aboriginal and Islander training program.

The Bachelor of Arts program is conducted over three years and commences each February. The program is vocationally based and professionally oriented. Most units in the curriculum combine theory, analysis and practical experience. Teaching methods include lectures, seminars, the viewing and analysis of work, written assignments and practical exercises. Production activity is augmented by studies of the screen and of the Australian and international film and television industry industries. Specializations are offered in cinematography, directing, editing, producing, production design, scriptwriting, and sound.

The School's teaching staff is drawn from experienced industry practitioners. Visiting film and television makers, critics, lecturers and other specialists also contribute to the teaching program and the assessment of students' work. The school has strong links with the film and broadcasting industries. Course development at AFTRS is strongly influenced by the needs of the industry. Leading professionals are members of the School's governing council.

Productions by students and graduates have won over 300 national and international awards. Former students can be found in all areas of the industry, including commercial radio and television, in government and independent production houses, and in arts and media administration.

Ball State University

Address	*Music Engineering Technology Studios* *Muncie* *IN* *47306* *USA*
Phone	*317-285-5537* **FAX** *317-285-5401*
Director	*Cleve Scott* **Admission Contact** *Dr. M. Vincent*
Program Founded	*1983* **School Type** *University*
Program Offered	*Music Engineering Technology*
Degrees Offered	*Bachelor of Music*
Program Length	*4 years*
Estimated Tuition	*$24,000*
Main Emphasis	*Electronic Music* **Program is** *Technical*
Accreditations	*NASM*
Number of Studios	*5* **Is school non-profit?** *Yes*
Types of Studios	○ *Acoustic Research* ◉ *Electronic Music* ○ *Film/Foley* ○ *Radio* ◉ *Audio Research* ◉ *Music Recording* ○ *Television* ○ *Video*
Types of Recording	◉ *Analog Multitrack* ◉ *MIDI Sequencing* ○ *Video* ○ *Digital Multitrack* ◉ *DAW* ○ *Film*
Other Resources	○ *Professional Studios* ○ *Television Stations* ○ *Radio Stations* ○ *Theater Tech Dept*
Class Size Lecture	*20* **Class Size Lab** *varies*
Assistance	○ *Housing* ◉ *Scholarships* ◉ *Internships* ◉ *Financial Aid* ◉ *Work-study* ◉ *Job Placement*
Admission Policy	*Highly selective* **Language** *English*
Prerequisites	◉ *High School Diploma* ◉ *SAT-ACT* ◉ *Music Audition*
Industry Affiliations	○ *AES* ○ *ASA* ○ *NACB* ○ *NARAS* ○ *SMPTE* ○ *NAMBI* ○ *APRS* ○ *NAB* ○ *NAMM* ○ *SBE* ○ *SPARS* ○ *MEIEA*
Fulltime Faculty	*10* **Parttime Faculty** *2*
Faculty Awards	*Indiana Arts Commission*
Program Awards	*National Review 1987*
Student Awards	*Presser Award, Young Artists Award*
Research areas	*Sound Synthesis & Performance, Computer Applications, Hardware Design*

Ball State University

The Music Engineering Technology (MET) curriculum is one of the most innovative programs at Ball State University, requiring a major in MET, a minor in applied physics/electronics, performance competency on an instrument, and composition skills. The combination of a major in music and a minor in applied physics is modeled after the European Tonmeister curriculum. The MET program emphasizes the art of music while appropriating science and technology as foundational and fundamental to the comprehension of music. In this context the music engineer might well be compared to an architect in that the music engineer is one who designs systems for the composition, performance, telecommunication, and recording of musical information.

Coursework is divided into three basic areas: courses in music engineering technology, courses in applied physics and electronics, and courses from the liberal arts curriculum. This pattern serves to produce four years of integrated experience, including individual accomplishments in composition, sound synthesis, computer applications, recording, MIDI and SMPTE communications, calculus-based physics and electronics, and a broad spectrum of general education courses to complement the individual's interests. Students are expected to accomplish junior standing on their major instrument, senior standing in composition, and a minimum of four semesters of large ensemble participation. Participation in small ensembles, jazz ensembles, University Singers, marching band, and new music ensembles is available by audition.

This is a challenging four-year program that focuses on both theoretical and practical experience. Students participate in a variety of extracurricular projects including field trips and lectures by experts in the music industry. Students in the MET program are eligible, through competition, for employment as engineers with Central Recording Services, University Singers, and the MET Studio.

This program is for the above-average student who is disciplined and self-directed. It is for the energetic and the industrious, requiring long hours of hard work and dedication to the goal of excellence in music. There is a conscious effort throughout the curriculum to individualize and specialize projects and assignments. This contributes to the development of a program that balances academic thinking with personal investigation and experimentation toward curricular goals of the student. The MET curriculum is one of dynamic learning, one that places as much emphasis on the creative endeavor as on the traditional learning experience.

Belmont University

Address *School of Business*
Nashville *TN*
37212 *USA*

Phone *615-386-4504* **FAX** *615-386-4516*

Director *Robert Mulloy* **Admission Contact** *Robert Mulloy*

Program Founded *1972* **School Type** *University*

Program Offered *Music Business*

Degrees Offered *Bachelor of Business Administration*

Program Length *4 years*

Estimated Tuition *$40,000*

Main Emphasis *Music Business* **Program is** *Semi-technical*

Accreditations *SACCS*

Number of Studios *2* **Is school non-profit?** *Yes*

Types of Studios ○ *Acoustic Research* ● *Electronic Music* ○ *Film/Foley* ○ *Radio*
○ *Audio Research* ● *Music Recording* ○ *Television* ○ *Video*

Types of Recording ● *Analog Multitrack* ● *MIDI Sequencing* ○ *Video*
● *Digital Multitrack* ○ *DAW* ○ *Film*

Other Resources ○ *Professional Studios* ○ *Television Stations*
○ *Radio Stations* ○ *Theater Tech Dept*

Class Size Lecture *30* **Class Size Lab** *12*

Assistance ○ *Housing* ● *Scholarships* ● *Internships*
● *Financial Aid* ● *Work-study* ● *Job Placement*

Admission Policy *Highly selective* **Language** *English*

Prerequisites ● *High School Diploma* ● *SAT-ACT* ○ *Music Audition*

Industry Affiliations ● *AES* ○ *ASA* ○ *NACB* ● *NARAS* ○ *SMPTE* ○ *NAMBI*
○ *APRS* ○ *NAB* ● *NAMM* ○ *SBE* ● *SPARS* ● *MEIEA*

Fulltime Faculty *6* **Parttime Faculty** *12*

Faculty Awards

Program Awards

Student Awards

Research areas

Belmont University

Belmont University offers a Music Business program that combines classroom experience with real world applications. After careful counsel with members of the Nashville music industry, Belmont pioneered a program in 1972 to prepare entry-level positions within the music industry. Celebrating its 20th anniversary in 1993, Belmont's Music Business program has created a network of over 600 graduates in the national as well as the local music industry. Graduates own management companies and recording studios and fill executive positions in major publishing companies and other industry-related businesses. While the emphasis of the program is business, some highly-successfully performing artists have Music Business degrees from Belmont.

The program constantly evaluates and implements curriculums which teach cutting edge management skills and technology. The full-time faculty and staff of music business professionals are complemented by adjunct professors who work full-time in the music industry.

Students who choose Music Business as a major are challenged to become young business executives. They realize the need to continually grow through the experience of internships and the cultivation of industry contacts. Belmont's active internship program provides opportunities for students to work in virtually any music industry company in Nashville and other cities in the United States. Students are encouraged to use their own initiatives to refine their career path.

The Center for Music Business houses and provides facilities reflecting the advances in the music industry. The academic resource center features: two state-of-the-art recording studios and control rooms, two additional studios, four isolation booths, a MIDI pre/post-production room, six practice/writing rooms, an all-purpose classroom, a Music Technology classroom with ten computer-based workstations, an engineering repair shop, and a fiber-optic link with the entire campus to enable on-site remote recording.

Berklee College of Music- MP & E

Address	*1140 Boylston Street* *Boston* *02215*	*MA* *USA*	
Phone	*617-266-1400*	**FAX**	*617-247-6878*
Director	*Don Puluse*	**Admission Contact**	*Steve Lipman*
Program Founded	*1983*	**School Type**	*University*
Program Offered	*Music Production & Engineering*		
Degrees Offered	*Bachelor of Music, Professional Diploma, Certificate*		
Program Length	*4 years, 2 years*		
Estimated Tuition	*$10,000 per year*		
Main Emphasis	*Music Recording*	**Program is**	*Technical*
Accreditations	*NEASC*		
Number of Studios	*10*	**Is school non-profit?**	*Yes*
Types of Studios	○ *Acoustic Research* ◉ *Electronic Music* ○ *Film/Foley* ○ *Radio* ○ *Audio Research* ◉ *Music Recording* ○ *Television* ○ *Video*		
Types of Recording	◉ *Analog Multitrack* ◉ *MIDI Sequencing* ○ *Video* ◉ *Digital Multitrack* ◉ *DAW* ○ *Film*		
Other Resources	○ *Professional Studios* ○ *Television Stations* ○ *Radio Stations* ○ *Theater Tech Dept*		
Class Size Lecture	*20*	**Class Size Lab**	*10*
Assistance	◉ *Housing* ◉ *Scholarships* ◉ *Internships* ◉ *Financial Aid* ◉ *Work-study* ◉ *Job Placement*		
Admission Policy	*Highly selective*	**Language**	*English*
Prerequisites	◉ *High School Diploma* ◉ *SAT-ACT* ◉ *Music Audition*		
Industry Affiliations	◉ *AES* ○ *ASA* ○ *NACB* ◉ *NARAS* ○ *SMPTE* ○ *NAMBI* ○ *APRS* ○ *NAB* ○ *NAMM* ○ *SBE* ◉ *SPARS* ◉ *MEIEA*		
Fulltime Faculty	*9*	**Parttime Faculty**	*20*
Faculty Awards			
Program Awards	*TEC Awards: 1985, 1986, 1987, 1992*		
Student Awards	*NARAS Student Music Award*		
Research areas			

Berklee College of Music- MP & E

The Music Production and Engineering (MP&E) major at Berklee College of Music is one of the world's leading programs in music recording. The program helps to mold musicians into recording professionals by drawing upon Berklee's strengths: a broad array of music technology, comprehensive recording facilities, knowledgeable and experienced faculty, and unequaled numbers of talented musicians. Berklee's goal is to graduate producers and engineers who understand the creative as well as the technical aspects of music and music recording.

The four-year MP&E program incorporates music, music technology, and music business in a comprehensive curriculum. Berklee students develop a competitive edge by first establishing a solid musical foundation through instrumental training and courses in harmony, arranging, ear training, music notation, and music history. MP&E students then develop and refine their technical and musical knowledge in Berklee's recording studio complex, which includes an array of multitrack studios, post-production/editing suites, synthesis labs, and MIDI-equipped ensemble/recital rooms.

MP&E students arriving at Berklee for the first time enter a microcosm of the music industry with all of its music, technical, and commercial considerations. With more than 2,500 students and 300 faculty members at their disposal- representing an eclectic mix of performers, composers, and arrangers, Berklee's fledgling producers and engineers have many opportunities to work on a one-to-one basis with artists on the cutting edge of contemporary music. Musical genres explored by Berklee performing ensembles include jazz, pop, rock, funk, and gospel.

Students in the MP&E program are provided with round-the-clock access to 10 recording studios and are assigned a series of production projects for which they have sole responsibility, from the conception of the music to the final mix of the master recording. While faculty members and experienced studio supervisors are available for assistance, students alone conduct all the engineering and production tasks. Each year, a faculty panel chooses the best MP&E student production projects for inclusion on the program's annual CD release. The National Academy of Recording Arts & Sciences granted a Student Music Award to a group of Berklee students who produced a track on the Studio Production Projects 1991 CD. The program has also won several TEC awards for Outstanding Institutional Achievement from MIX magazine.

Students have a chance to learn from some of the industry's acclaimed recording professionals through the college's Visiting Artist Series. Master classes and lectures have been given by producer/artist Nile Rodgers, mastering engineer Bob Ludwig, producer Arif Mardin, and producer/arranger Barry Eastmond. Program director Don Puluse, formerly a recording engineer with CBS records, has earned several gold and platinum records for his work on albums by Janis Joplin, Bob Dylan, and Chicago. Some renown Berklee alumni include multi-award winning composer, arranger and producer Quincy Jones, VP and music director of Atlantic Records Arif Mardin, and engineer/producer Roger Talkov.

Berkleee College of Music- Synthesis

Address	*1140 Boylston Street*		
	Boston	*MA*	
	02215	*USA*	
Phone	*617-266-1400*	**FAX**	*617-536-2632*
Director	*Dennis Thurmond*	**Admission Contact**	*Steve Lipman*
Program Founded	*1985*	**School Type**	*University*
Program Offered	*Music Synthesis*		
Degrees Offered	*Bachelor of Music, Professional Diploma*		
Program Length	*4 years*		
Estimated Tuition	*$10,000 per year*		
Main Emphasis	*Electronic Music*	**Program is**	*Technical*
Accreditations	*NEASC*		
Number of Studios	*6*	**Is school non-profit?**	*Yes*
Types of Studios	○ *Acoustic Research* ◉ *Electronic Music* ○ *Film/Foley* ○ *Radio* ○ *Audio Research* ◉ *Music Recording* ○ *Television* ○ *Video*		
Types of Recording	◉ *Analog Multitrack* ◉ *MIDI Sequencing* ○ *Video* ◉ *Digital Multitrack* ◉ *DAW* ○ *Film*		
Other Resources	○ *Professional Studios* ○ *Television Stations* ○ *Radio Stations* ○ *Theater Tech Dept*		
Class Size Lecture	*6-80*	**Class Size Lab**	*12*
Assistance	◉ *Housing* ◉ *Scholarships* ○ *Internships* ◉ *Financial Aid* ◉ *Work-study* ◉ *Job Placement*		
Admission Policy	*Selective*	**Language**	*English*
Prerequisites	◉ *High School Diploma* ◉ *SAT-ACT* ◉ *Music Audition*		
Industry Affiliations	◉ *AES* ○ *ASA* ○ *NACB* ◉ *NARAS* ○ *SMPTE* ○ *NAMBI* ○ *APRS* ○ *NAB* ○ *NAMM* ○ *SBE* ◉ *SPARS* ◉ *MEIEA*		
Fulltime Faculty	*6*	**Parttime Faculty**	
Faculty Awards			
Program Awards			
Student Awards			
Research areas			

Berklee College of Music- Synthesis

Berklee's Music Synthesis major is designed to keep students on top of the technological advances in the music industry. Through a series of courses in subtractive synthesis, FM digital synthesis, hard-disk recording and editing systems, and digital sampling techniques, students complete a number of creative projects in each area while learning the conceptual bases of these applications.

Music Synthesis majors, by choosing a wide range of elective courses, can specialize in any of three directions: performance, production, or sound design. Students interested in live performance can participate in ensembles specifically geared for synthesists. Ensembles rehearse and give recitals in state-of-the-art performance facilities equipped for the performing specialist's needs. The production track allows students to concentrate on synthesis production course work including composition, orchestration, advanced sequencing applications, and hard-disk direct-to-digital recording and editing. Students produce completed projects on DAT. Sound design students can explore electronic sound creation and manipulation in course work that includes advanced sound design for musical and sound effect applications. Students create a portfolio of original sounds for a variety of musical situations and instrumental combinations.

Berklee has created three synthesis classroom/labs with a full array of synthesizers and supporting equipment; two synthesizer-equipped rehearsal rooms; a 75-seat MIDI performance hall; and a new 40-workstation multimedia computing laboratory. Each classroom/lab is designed for a specific type of music synthesis and contains individualized student workstations as well as a teaching station and sound system. The ensemble rooms are equipped with synthesizers, computers, and sound reinforcement gear for multi-keyboard/controller ensemble rehearsals.

The Subtractive Lab contains twelve complete Macintosh-based student workstations, enabling interaction with synthesizers, mixers, effects units, and other hardware. The Digital Lab also has twelve Mac-based workstations, enabling students to complete CD-quality projects. The Advanced Systems Lab has six workstations equipped with a combination of software and hardware that encompasses a variety of synthesis and multitrack hard-disk technology.

Music Synthesis majors benefit from Berklee's professional recording facilities and opportunities to work with hundreds of talented musicians as well as experienced and talented faculty. Many Music Synthesis majors pursue double majors with disciplines such as film scoring, commercial arranging, performance, songwriting, and composition.

Members of Berklee's full-time Music Synthesis faculty share a comprehensive and varied background in music, technology, and education, including extensive professional experience in recording and production, instrument design, performance, composition, and theater. A number of the music industry's prominent synthesists received their start at Berklee, including Joseph Zawinul, Rob Mounsey, Jan Hammer, and Jeff Lorber.

California Institute of the Arts, Music

Address 24700 McBean Parkway
Valencia CA
91355 USA

Phone 805-255-1050 **FAX** 805-254-8352

Director David Rosenboom **Admission Contact** David Rosenboom

Program Founded 1971 **School Type** University

Program Offered Music Technology and New Media

Degrees Offered BFA, MFA, Certificate, Advanced Certificate

Program Length 4 years, 2 years

Estimated Tuition $13,850 per year

Main Emphasis Electronic Music **Program is** Semi-technical

Accreditations WASC, NASM

Number of Studios 5 **Is school non-profit?** Yes

Types of Studios ○ Acoustic Research ○ Electronic Music ○ Film/Foley ○ Radio
○ Audio Research ○ Music Recording ○ Television ○ Video

Types of Recording ◉ Analog Multitrack ◉ MIDI Sequencing ◉ Video
○ Digital Multitrack ◉ DAW ◉ Film

Other Resources ◉ Professional Studios ○ Television Stations
○ Radio Stations ◉ Theater Tech Dept

Class Size Lecture 10 **Class Size Lab** 8

Assistance ○ Housing ◉ Scholarships ○ Internships
◉ Financial Aid ◉ Work-study ○ Job Placement

Admission Policy Selective **Language** English

Prerequisites ◉ High School Diploma ○ SAT-ACT ◉ Music Audition

Industry Affiliations ○ AES ○ ASA ○ NACB ○ NARAS ○ SMPTE ○ NAMBI
○ APRS ○ NAB ○ NAMM ○ SBE ○ SPARS ○ MEIEA

Fulltime Faculty 8 **Parttime Faculty** 3

Faculty Awards

Program Awards

Student Awards

Research areas Interactive computer media in performing arts

California Institute of the Arts, Music

Building on the school's established leadership in composition and performance with new musical languages, the Music Technology and New Media pathway in the Music Arts Program offers students an opportunity to concentrate on developing expertise in new media technology while acquiring a solid background in the fundamentals of musical practice and experiencing CalArts' global perspective on music making. CalArts active and creative performance environment offers nearly limitless opportunities to work with performers and composers in production projects, providing a variety of hands-on learning experiences. In addition, the presence of schools in all the arts within the Institute encourages substantive, interdisciplinary collaboration. CalArts faculty includes many innovators responsible for major developments in a number of fields. The presence of the Center for Experiments in Art, Information and Technology (CEAIT) helps create an atmosphere for invention and experimentation. Within the rapidly expanding world of multi-media, students who possess technical skills and a deep understanding of the musical arts may find themselves ideally equipped for careers of the future.

The expansion of new arts media made possible by digital computer technology has created art media fields extending far beyond the role of the traditional recording engineer. This program emphasizes those new directions that have emerged from music-making, including interactive music software, integrated multi-media, MIDI systems, computer music techniques, music printing and DTP, CD-ROM technology, algorithmic music, live performance with interactive instruments, sound design, new directions in sound synthesis and live processing, digital signal processing, concert production, perception of sound, sound recording, and experiments with communication networks.

The music composition department at CalArts benefits from strong programs in jazz and African-American music. Annual festivals and concerts provide students with exposure and performance opportunities in a variety of contemporary music ensembles. CalArts also features a visiting artists program that regularly brings outstanding lecturers to the school.

California Recording Institute

Address *970 O'Brien Drive*
Menlo Park *CA*
94025 *USA*

Phone *415-324-0464* **FAX**

Director *David Gibson* **Admission Contact** *Donna Compton*

Program Founded *1981* **School Type** *Trade School*

Program Offered *Recording Arts and Technology*

Degrees Offered *Certificate*

Program Length *1 year*

Estimated Tuition *$6,000*

Main Emphasis *Audio Engineering* **Program is** *Semi-technical*

Accreditations

Number of Studios *6* **Is school non-profit?** *No*

Types of Studios O *Acoustic Research* ◉ *Electronic Music* O *Film/Foley* O *Radio*
O *Audio Research* ◉ *Music Recording* O *Television* O *Video*

Types of Recording ◉ *Analog Multitrack* ◉ *MIDI Sequencing* ◉ *Video*
◉ *Digital Multitrack* ◉ *DAW* ◉ *Film*

Other Resources ◉ *Professional Studios* ◉ *Television Stations*
O *Radio Stations* O *Theater Tech Dept*

Class Size Lecture *13* **Class Size Lab** *6*

Assistance ◉ *Housing* O *Scholarships* ◉ *Internships*
◉ *Financial Aid* O *Work-study* ◉ *Job Placement*

Admission Policy **Language** *English*

Prerequisites O *High School Diploma* O *SAT-ACT* O *Music Audition*

Industry Affiliations ◉ *AES* O *ASA* O *NACB* ◉ *NARAS* O *SMPTE* O *NAMBI*
O *APRS* O *NAB* ◉ *NAMM* O *SBE* ◉ *SPARS* O *MEIEA*

Fulltime Faculty *7* **Parttime Faculty** *4*

Faculty Awards

Program Awards

Student Awards

Research areas *Visual Sound Display for Mix Control*

California Recording Institute

The California Recording Institute offers a one-year program in Recording Arts and Technology. The program covers the skills and techniques required of a professional engineer and provides the experience to master them. The program is the first to use visual representations of imaging, the placement of sounds between the speakers, to describe the process of mixing. Classes are small and students receive as much personal support and scrutiny as desired.

Practical training is provided at Music Annex Recording Studios, a professional recording facility. Classes are held in five different studios to give students a good perspective of the differences between mixing consoles and control rooms. Students are also instructed in the use of automated consoles. The program emphasizes extensive hands-on experience.

Classes are one night a week and one night every other week with an occasional Saturday. The course consists of 358 hours of training, approximately one-third devoted to lecture and two-thirds devoted to hands-on recording time. Classes include: The Art and Technology of Production, Hands-on Recording/Mixing, Equipment Maintenance, MIDI, Digital Recording, Studio Computers, Hard Disk Recording, Record Companies and the Music Business, Sound Reinforcement, and Introduction to Television Production. All classes are also offered on an individual basis.

For those interested in an introductory class, a six-week Recording/Mixing Class is offered. The course is aimed at beginners, musicians, home engineering, and experienced engineers wishing to improve their understanding of modern recording techniques. The course meets one night a week and is limited to 15 students.

California State University at Chico

Address *Music Department*
Chico *CA*
95929-0805 *USA*

Phone *916-898-5152* **FAX** *916-898-4082*

Director *Raymond Barker* **Admission Contact** *Registrar*

Program Founded *1987* **School Type** *University*

Program Offered *Option in Recording Arts*

Degrees Offered *Bachelor of Arts in Music*

Program Length *4 years*

Estimated Tuition *$4,400*

Main Emphasis *Audio Engineering* **Program is** *Semi-technical*

Accreditations *NASM*

Number of Studios *3* **Is school non-profit?** *Yes*

Types of Studios ○ *Acoustic Research* ◉ *Electronic Music* ◉ *Film/Foley* ○ *Radio*
○ *Audio Research* ◉ *Music Recording* ○ *Television* ○ *Video*

Types of Recording ◉ *Analog Multitrack* ◉ *MIDI Sequencing* ○ *Video*
◉ *Digital Multitrack* ◉ *DAW* ◉ *Film*

Other Resources ○ *Professional Studios* ○ *Television Stations*
○ *Radio Stations* ◉ *Theater Tech Dept*
Instructional Media Center

Class Size Lecture *24* **Class Size Lab** *12*

Assistance ○ *Housing* ◉ *Scholarships* ◉ *Internships*
◉ *Financial Aid* ○ *Work-study* ◉ *Job Placement*

Admission Policy *Open* **Language** *English*

Prerequisites ◉ *High School Diploma* ◉ *SAT-ACT* ○ *Music Audition*

Industry Affiliations ◉ *AES* ○ *ASA* ○ *NACB* ○ *NARAS* ○ *SMPTE* ○ *NAMBI*
○ *APRS* ○ *NAB* ○ *NAMM* ○ *SBE* ○ *SPARS* ○ *MEIEA*

Fulltime Faculty *2* **Parttime Faculty** *1*

Faculty Awards *Gold/Platinum Records, Teaching Awards*

Program Awards *Three nationally-released CD projects by students*

Student Awards

Research areas

California State University at Chico

The Department of Music at California State University, Chico, offers two programs in Recording Arts: the Bachelor of Arts in Music with an Option in Recording Arts and the Minor in Recording Arts. These programs are offered in recording facilities located in the west wing of the University's Performing Arts Center. The facilities include a multitrack control room, a performance studio, and an electronic music studio.

The Option in Recording Arts provides a curriculum for students wishing to seek employment in fields combining music and technology. A music major in the Recording Arts Option takes courses in music history, music theory, composition with electronic media, audio recording, basic electronics with electives in audio for video and the music industry. Completion of a B.A. in this program offers music majors enhanced employment opportunities in technical areas--areas for example in the recording industry (recording engineer, director, producer, editor, music coordinator, etc.) and in music synthesis (electronic musical instruments—composition, performance, teaching, retail sales, etc.). Employment opportunities can be further improved by pairing the Option in Recording Arts with a minor in an area such as business, telecommunications, engineering, computer science, or electronics. The B.A. with an Option in Recording Arts is also preparation for a Master's degree for those wishing to continue their educations.

As a corollary to the Option, the Minor in Recording Arts assists students in other departments and disciplines with becoming more employable. It provides courses in music fundamentals, music appreciation, composition with electronic media, audio recording, and electronics. The Minor in Recording Arts will appeal to students majoring in areas such as Telecommunications, Engineering, Computer Science, Industrial Technology, Business, Theater Arts, Dance, Education and Physical Science.

The Music Department at California State University, Chico, with performing groups as diverse as the Chico Symphony Orchestra and the Jazz Ensemble, chamber quartets, operas and musicals, offers an ideal hands-on training laboratory for students in recording arts. Audio recording courses provide the opportunity to produce and record pop, jazz, and classical productions. An annual Electronic Music and Media Concert showcases student compositions created in the composition with electronic media courses. The program's new recording studio is linked directly to the university's National Public Radio affiliated station, KCHO-FM, with the capability of live broadcasts, and fosters working relationships among students, faculty, and technicians in the performing arts, telecommunications, audio, and instructional media.

Campus AV- Professional Audio/Video Training

Address *7 Miller Close, Offord Darcy*
Huntingdon *Cambridgeshire*
PE18 9SB *UK*

Phone *0480 812201* **FAX** *0480 812280*

Director *David Pope* **Admission Contact** *David Pope*

Program Founded *1989* **School Type** *Workshop*

Program Offered *Sound Engineering*

Degrees Offered *Diploma*

Program Length *12 weeks*

Estimated Tuition *£2,750 + VAT*

Main Emphasis *Music Recording* **Program is** *Semi-technical*

Accreditations

Number of Studios *5* **Is school non-profit?** *Yes*

Types of Studios
O *Acoustic Research* ◉ *Electronic Music* O *Film/Foley* O *Radio*
O *Audio Research* ◉ *Music Recording* O *Television* O *Video*

Types of Recording
◉ *Analog Multitrack* ◉ *MIDI Sequencing* ◉ *Video*
O *Digital Multitrack* O *DAW* O *Film*

Other Resources
◉ *Professional Studios* O *Television Stations*
O *Radio Stations* O *Theater Tech Dept*
PA Company, Music Factory

Class Size Lecture *12* **Class Size Lab** *4*

Assistance
◉ *Housing* O *Scholarships* O *Internships*
◉ *Financial Aid* O *Work-study* O *Job Placement*

Admission Policy *Selective* **Language** *English*

Prerequisites O *High School Diploma* O *SAT-ACT* O *Music Audition*

Industry Affiliations
O *AES* O *ASA* O *NACB* O *NARAS* O *SMPTE* O *NAMBI*
O *APRS* O *NAB* O *NAMM* O *SBE* O *SPARS* O *MEIEA*

Fulltime Faculty *4* **Parttime Faculty** *20*

Faculty Awards

Program Awards

Student Awards

Research areas

Campus AV- Professional Audio/Video Training

The Campus AV's Sound Engineering course runs for a full three months- July, August and September. The course is designed for those seeking a career in sound recording, concentrating principally on the multitrack music recording business. An overview is provided for other areas such as sound reinforcement and sound dubbing for pictures. The objective is to provide a solid grounding in sound recording and post-production techniques thus allowing students to choose a career direction most appropriate to their own interests.

The three month course is structured into three one-month parts. The first month is an introductory course that seeks to lay a foundation for a thorough understanding of sound principles. An overview of the complete process is presented where the purpose of each piece of equipment is examined as well as the role of each person involved in the business. An introduction to video and film is also covered. The second month is an intermediate-level designed for those who already have a reasonably good understanding of the principles involved. A high level of practical and operational hands-on training is provided. Students are taught how to operate the whole range of studio equipment and are tested on their aptitude. Guest lecturers from leading companies and studios go into greater detail on such topics as microphone techniques, acoustics, and MIDI programming. In the final month, students work on individual projects that utilize all the techniques covered in the previous two months. Opportunities are provided for each student to record and mix individual projects. Advanced digital audio equipment is also studied in detail.

Campus AV offers additional short courses (2-3 day) in music recording, live DJ mixing, live sound, digital audio technology, and video. They are designed with continuing education in mind and cost from £199 to £299 + VAT. To ensure individual attention, class sizes are limited to 20 students. For practical experiences the class is divided into small groups of five persons or less.

The courses are all taught at The University of Surrey, benefiting from their high quality audio facilities and Tonmeister degree program. Students can take advantage of the University's extensive research library and study facilities. The program's main lecturers, David Pope, John Watkinson, Francis Rumsey, and Ken Blair, are respected members of the professional audio industry. A regular series of guest lecturers also provide practical and essential current views of their sector of the industry.

Cayuga Community College

Address	*197 Franklin Street*		
	Auburn	*NY*	
	13021	*USA*	
Phone	*315-255-1743*	**FAX**	*315-255-2050*
Director	*Steven Keeler*	**Admission Contact**	*Patricia Powers-Burdick*
Program Founded	*1971*	**School Type**	*Community College*
Program Offered	*Audio/Radio Production*		
Degrees Offered	*Associate in Applied Science*		
Program Length	*2 years*		
Estimated Tuition	*$3,700*		
Main Emphasis	*Audio Engineering*	**Program is**	*Semi-technical*
Accreditations	*Middle States, SBE*		
Number of Studios	*7*	**Is school non-profit?**	*Yes*

Types of Studios
- ○ *Acoustic Research* ○ *Electronic Music* ○ *Film/Foley* ◉ *Radio*
- ○ *Audio Research* ◉ *Music Recording* ◉ *Television* ◉ *Video*

Types of Recording
- ◉ *Analog Multitrack* ○ *MIDI Sequencing* ◉ *Video*
- ○ *Digital Multitrack* ○ *DAW* ○ *Film*

Other Resources
- ○ *Professional Studios* ○ *Television Stations*
- ◉ *Radio Stations* ◉ *Theater Tech Dept*

Class Size Lecture	*32*	**Class Size Lab**	*14*

Assistance
- ◉ *Housing* ◉ *Scholarships* ◉ *Internships*
- ◉ *Financial Aid* ◉ *Work-study* ◉ *Job Placement*

Admission Policy	*Open*	**Language**	*English*

Prerequisites
- ◉ *High School Diploma* ◉ *SAT-ACT* ○ *Music Audition*

Industry Affiliations
- ○ *AES* ○ *ASA* ○ *NACB* ○ *NARAS* ○ *SMPTE* ○ *NAMBI*
- ○ *APRS* ◉ *NAB* ○ *NAMM* ◉ *SBE* ○ *SPARS* ○ *MEIEA*

Fulltime Faculty	*1*	**Parttime Faculty**	*8*

Faculty Awards

Program Awards

Student Awards

Research areas

Cayuga Community College

Cayuga Community College offers a two-year Associate in Applied Science degree for students interested in becoming audio production specialists for radio, television, video, and the recording industry. The program prepares students for entry-level positions. Upon completion of the degree, students will be able to perform such audio production functions as spot recording and audio programming, mixing, re-recording, splicing, dubbing and overdubbing, on-location sound reinforcement, and site planning.

The course work and internship for the program can be completed in a four-semester sequence. The academic content of the courses emphasizes training in and understanding of the practices and business aspects of the radio, television, and recording industries. This includes planning, producing, and executing productions, as well as using various production equipment. Students attain the skills necessary to effectively perform audio production functions.

In the first semester, students are introduced to the broadcasting and recording industries. Emphasis is placed on the responsibilities of each staff member in a production facility. Students learn basic production techniques. During the second semester, students learn to plan and execute radio and television productions. Areas covered include advanced audio techniques, sound reinforcement, recording studio techniques, and acoustic analysis.

In the third semester, students gain hands-on equipment operations experience by producing several studio and remote television programs for cablecast. Emphasis is placed on multitrack recording technology, including overdubbing, remixing, and signal processing. The fourth semester is devoted to providing students with extensive experience in the techniques and operations of audio/radio production. Students practice and refine their understanding of telecommunications both at the campus studios and in an off-campus facility during their internship.

Cayuga Community College also offers a special workshop, Broadcasting: The British Experience, through their International Education Program. This study-travel course offers a comprehensive view of the broadcasting industry in Great Britain. The course consists of an intensive two-week series of tours, lectures, backstage and audience observation of television productions, and discussions.

Center for Media Arts at Mercy College

Address *226 West 26th Street*
New York *NY*
10001 *USA*

Phone *800-262-2297* **FAX** *212-463-0746*

Director *David Sanders* **Admission Contact** *Joy Colleli*

Program Founded *1984* **School Type** *University*

Program Offered *Basic Audio Recording Techniques with Specializations in Audio & Music*

Degrees Offered *Bachelor of Arts, Associate of Arts, Certificate*

Program Length *6 months- 4 years*

Estimated Tuition *varies*

Main Emphasis *Music Recording* **Program is** *Semi-technical*

Accreditations *MSAC, NYS Board of Regents*

Number of Studios *7* **Is school non-profit?** *Yes*

Types of Studios
○ *Acoustic Research* ◉ *Electronic Music* ○ *Film/Foley* ◉ *Radio*
○ *Audio Research* ◉ *Music Recording* ◉ *Television* ◉ *Video*

Types of Recording
◉ *Analog Multitrack* ◉ *MIDI Sequencing* ◉ *Video*
○ *Digital Multitrack* ◉ *DAW* ○ *Film*

Other Resources
○ *Professional Studios* ○ *Television Stations*
◉ *Radio Stations* ○ *Theater Tech Dept*

Class Size Lecture *18* **Class Size Lab** *12*

Assistance
○ *Housing* ○ *Scholarships* ◉ *Internships*
◉ *Financial Aid* ○ *Work-study* ◉ *Job Placement*

Admission Policy *Open* **Language** *English*

Prerequisites ◉ *High School Diploma* ○ *SAT-ACT* ○ *Music Audition*

Industry Affiliations
◉ *AES* ○ *ASA* ○ *NACB* ○ *NARAS* ○ *SMPTE* ○ *NAMBI*
○ *APRS* ○ *NAB* ○ *NAMM* ○ *SBE* ◉ *SPARS* ○ *MEIEA*

Fulltime Faculty *3* **Parttime Faculty** *25*

Faculty Awards

Program Awards

Student Awards

Research areas

Center for Media Arts at Mercy College

The Center for Media Arts' Basic Audio Recording Techniques (BART) program serves as the core curriculum for students who may later elect to register for one of three specializations: Audio Engineer, MIDI/Synthesizer Specialist, and Music Production. The BART core curriculum trains students in the fundamental skills necessary for all audio-related careers. Students gain a solid foundation of understanding in the following areas: acoustics, mixing, editing, multitrack recording, music industry operations, computer/synthesizer music production techniques, MIDI, signal processing, audio post-production for video, and digital editing techniques.

The Audio Engineer specialty trains students for careers in various aspects of professional audio. In addition to covering equipment operation and management it also provides technical knowledge aimed at enhancing the careers of recording artists, producers and composers. The course focuses on digital audio and the tapeless studio concept. Students spend nearly a month in a state-of-the-art digital control room, working with an advanced digital audio workstation and automated mixing consoles. Students sharpen the mixing, editing and post-production skills they began to develop in the core program. Topics such as basic studio maintenance practices and sound reinforcement are covered, providing the groundwork for the technical knowledge needed for success in the contemporary audio field. A more in-depth look at music industry operation and studio management gains students insight into the business side of the recording studio.

The Music Production curriculum provides training in all aspects of music production and places special emphasis on commercial music styles. Students graduating from the program will be equipped to compose, arrange and record music for the enhancement of their own recording careers or the careers of others. They gain insight into the workings of the music industry from the prospective of the producer. Finally, each student produces a high quality demonstration tape of their work in CMA's recording studio and MIDI lab. Because a significant portion of the curriculum deals with the composition and arrangement of contemporary music, a high level of musical competency is required for admittance to this program. In addition, students are required to complete 24 hours of weekend music theory classes while attending the BART core curriculum.

The MIDI/Synthesizer Specialist program trains students for a career as a synthesist/programmer for live or recorded music production. It also provides technical expertise to enhance the careers of recording artists, instrumentalists, producers and composers of music for films and advertising. Emphasis is placed on the use of personal computers in sound design and in sequencing techniques for music production and video sound tracks.

Career placement is offered after completion of the BART program, but students are strongly recommended to complete additional courses in one of the three specializations.

Cleveland Institute of Music

Address	*11021 East Blvd* *Cleveland* *44106*	*OH* *USA*	
Phone	*216-791-5000*	**FAX**	*216-791-3063*
Director	*Thomas Knab*	**Admission Contact**	*William Fay*
Program Founded	*1983*	**School Type**	*University*
Program Offered	*Audio Recording*		
Degrees Offered	*Bachelor of Music*		
Program Length	*4 years*		
Estimated Tuition	*$12,250 per year*		
Main Emphasis	*Music Recording*	**Program is**	*Semi-technical*
Accreditations	*NASM*		
Number of Studios	*5*	**Is school non-profit?**	*Yes*

Types of Studios
◉ *Acoustic Research* ◉ *Electronic Music* ◉ *Film/Foley* ◉ *Radio*
○ *Audio Research* ◉ *Music Recording* ○ *Television* ○ *Video*

Types of Recording
◉ *Analog Multitrack* ◉ *MIDI Sequencing* ○ *Video*
◉ *Digital Multitrack* ◉ *DAW* ○ *Film*

Other Resources
◉ *Professional Studios* ◉ *Television Stations*
◉ *Radio Stations* ○ *Theater Tech Dept*

Class Size Lecture *10* **Class Size Lab** *10*

Assistance
◉ *Housing* ◉ *Scholarships* ◉ *Internships*
◉ *Financial Aid* ◉ *Work-study* ○ *Job Placement*

Admission Policy *Highly selective* **Language** *English*

Prerequisites
◉ *High School Diploma* ◉ *SAT-ACT* ◉ *Music Audition*

Industry Affiliations
◉ *AES* ○ *ASA* ○ *NACB* ○ *NARAS* ○ *SMPTE* ○ *NAMBI*
○ *APRS* ○ *NAB* ○ *NAMM* ○ *SBE* ○ *SPARS* ○ *MEIEA*

Fulltime Faculty *4* **Parttime Faculty** *5*

Faculty Awards

Program Awards

Student Awards

Research areas *Acoustics*

Cleveland Institute of Music

The Cleveland Institute of Music Audio Recording program is designed for the classical musician who wishes to learn the fundamentals of audio/music production while pursuing an undergraduate college degree in a music conservatory setting. A Bachelor of Music degree in Audio Recording is offered through both major and double-major programs; a minor in electrical engineering is available through the Joint Music Program with Case Western Reserve University.

CIM's audio curriculum covers audio system design and operation, digital audio, advanced microphone and recording techniques for classical music ensembles, audio for video production and post-production, synthesis and electronic music, and multitrack recording techniques. An internship in a professional studio, independent projects, and four years of professional experience in the CIM Audio Service round out the student's experience.

Classical music recording studies are done under the guidance of adjunct faculty member Jack Renner, CEO and chief recording engineer of Telarc International. Acoustics presentations and experiments are with adjunct faculty member Dr. Peter D'Antonio, president of RPG Diffusor Systems. Top professionals in the audio field are regularly invited to CIM as guest lecturers.

Two student recording control rooms and a recording studio allow students plenty of hands-on time. Students have access to all studio equipment by their sophomore year, and assignments for practicing recording techniques are emphasized in each class. CIM's conservatory setting offers excellent large ensembles, chamber music ensembles, and other talented musicians for collaboration on recording projects.

The program accepts only 16 majors to assure personalized instruction, copious studio time, and support with internships and job searches. Students have pursued internships at such prominent places as the Aspen Music Festival, Tanglewood Music Festival, Interlochen, Blossom Music Festival, and Suma Recording to widen their audio and video experience, make professional contacts, and clarify their career goals.

Admission is based the results of music theory placement tests, general knowledge of Classical, Romantic, and Contemporary music, audio knowledge tests, personal interview, and high school grades. Audio recording applicants who are not planning to double major in performance must also demonstrate technical and musical accomplishment on an instrument. The audition repertoire is the same as required for an entrance audition on that instrument. Acceptable levels of performance ability will be determined by a member of that department and the head of the Audio Recording department.

Cogswell Polytechnical College

Address	*10420 Bubb Road*		
	Cupertino	*CA*	
	95014	*USA*	
Phone	*408-252-5550*	**FAX**	*408-253-2413*
Director	*Eric Peterson*	**Admission Contact**	*Paul Schreivogel*
Program Founded	*1988*	**School Type**	*University*
Program Offered	*Music Engineering Technology*		
Degrees Offered	*Bachelor of Science*		
Program Length	*8 trimesters*		
Estimated Tuition	*$26,400*		
Main Emphasis	*Audio Engineering*	**Program is**	*Technical*
Accreditations	*WASC, TAC/ABET*		
Number of Studios	*3*	**Is school non-profit?**	*Yes*

Types of Studios
- ○ *Acoustic Research*
- ◉ *Electronic Music*
- ○ *Film/Foley*
- ○ *Radio*
- ○ *Audio Research*
- ○ *Music Recording*
- ○ *Television*
- ◉ *Video*

Types of Recording
- ◉ *Analog Multitrack*
- ◉ *MIDI Sequencing*
- ◉ *Video*
- ○ *Digital Multitrack*
- ◉ *DAW*
- ○ *Film*

Other Resources
- ◉ *Professional Studios*
- ○ *Television Stations*
- ○ *Radio Stations*
- ○ *Theater Tech Dept*

Silicon Valley Industries

Class Size Lecture	*1-12*	**Class Size Lab**	*1-8*

Assistance
- ◉ *Housing*
- ○ *Scholarships*
- ○ *Internships*
- ◉ *Financial Aid*
- ◉ *Work-study*
- ◉ *Job Placement*

Admission Policy	*Selective*	**Language**	*English*

Prerequisites
- ◉ *High School Diploma*
- ◉ *SAT-ACT*
- ○ *Music Audition*

Industry Affiliations
- ◉ *AES*
- ○ *ASA*
- ○ *NACB*
- ○ *NARAS*
- ○ *SMPTE*
- ○ *NAMBI*
- ○ *APRS*
- ○ *NAB*
- ○ *NAMM*
- ○ *SBE*
- ○ *SPARS*
- ◉ *MEIEA*

Fulltime Faculty	*2*	**Parttime Faculty**	*6*

Faculty Awards

Program Awards

Student Awards

Research areas

Cogswell Polytechnical College

Cogswell Polytechnical College offers the only Bachelor of Science degree in Music Engineering Technology in the world and was selected as one of the nation's top engineering schools by US News & World Reports magazine in 1991. Their Music Engineering Technology program is aimed at students who are interested in blending their love for computers, music, math, and science into a program of music production, engineering internships, new product testing and evaluation, and media process and techniques. Hands-on lab experience begins from the first day of the program.

Student engineers create music and sound in Cogswell's MIDI labs and audio recording studios, design and test hardware in the electronic labs, and write and debug software in the software engineering labs. All graduates build and demonstrate working senior projects and prepare professional portfolios. Students have opportunities to obtain dual degrees in computer and electronic engineering technology as well as minor in software engineering. Elective classes in integrate media are available in the college's Computer and Video Imaging degree program.

The program is supported by teaching and lab facilities equipped with flagship stations from major music technology companies, third party developers, and Silicon Valley manufacturers. An on-site, professional multitrack audio recording facility and two additional student studios provide MIDI and digital audio systems for tape and digital recording, as well as audio for video post-production.

The program is supported by the Cogswell Music Technology Industrial Advisory Board, representing all the major Silicon Valley music technology companies. Annual meetings with the board ensure that Cogswell graduates are well prepared for careers in the computer, electronic, music, and integrated media fields. The school also sponsors an annual music technology expo that attracts the leaders of the electronic music and computer industry to the school for educational seminars and demonstrations of cutting edge technology. In addition, the program benefits from active student chapters of the AES and MEISA.

College for Recording Arts

Address *665 Harrison Street*
San Francisco *CA*
94107 *USA*

Phone *415-781-6306* **FAX** *415-781-0115*

Director *Leo De Gar Kulka* **Admission Contact** *Patricia Lowery*

Program Founded *1974* **School Type** *Trade School*

Program Offered *Recording/Music Business*

Degrees Offered *Diploma*

Program Length *1 year*

Estimated Tuition *$9,000*

Main Emphasis *Audio Engineering* **Program is** *Technical*

Accreditations *CCA, California Dept of Ed*

Number of Studios *2* **Is school non-profit?** *No*

Types of Studios O *Acoustic Research* O *Electronic Music* O *Film/Foley* ◉ *Radio*
O *Audio Research* ◉ *Music Recording* O *Television* O *Video*

Types of Recording ◉ *Analog Multitrack* ◉ *MIDI Sequencing* O *Video*
◉ *Digital Multitrack* O *DAW* O *Film*

Other Resources ◉ *Professional Studios* O *Television Stations*
◉ *Radio Stations* O *Theater Tech Dept*
DAW Mastering

Class Size Lecture *35* **Class Size Lab** *14*

Assistance ◉ *Housing* ◉ *Scholarships* ◉ *Internships*
◉ *Financial Aid* ◉ *Work-study* ◉ *Job Placement*

Admission Policy *Selective* **Language** *English*

Prerequisites ◉ *High School Diploma* O *SAT-ACT* O *Music Audition*

Industry Affiliations ◉ *AES* O *ASA* O *NACB* ◉ *NARAS* ◉ *SMPTE* O *NAMBI*
O *APRS* ◉ *NAB* O *NAMM* O *SBE* O *SPARS* O *MEIEA*

Fulltime Faculty *2* **Parttime Faculty** *11*

Faculty Awards

Program Awards

Student Awards

Research areas

College for Recording Arts

The College for Recording Arts' aim is to provide a specialized training ground for students interested in the recording and music industries. One of the oldest school's in existence, CRA was conceived as an environment for students to learn the relationship between the recording and music fields covering the spectrum from the conception of an idea to retail sales. The subjects presented include studio engineering, disc cutting, record production, record marketing, music publishing and promotion, as well as a thorough examination of the economic and business aspects of the industry. New courses, such as audio/video post-production and studio maintenance, are also included.

All courses are supervised by professionals who are successfully active in the industry. They provide hands-on experience in a professional environment. The recording facility at the school has longed served as a professional recording studio and currently serves an independent record label. The studio is commercially active, allowing students to observe and participate in recording sessions to expose them to the challenges and pressures of professional recording work.

The diploma program consists of three 14-week semesters with thirteen required courses. Optional classes in specialized subjects can be used for advanced study to compliment the fundamental diploma core of classes. Included in these optional courses are sound reinforcement, electronic music, disc mastering and CD preparation, radio drama and commercial production, and other specialized seminars as instructors are available.

A limited program of individual study is available at the College for students not intending to take the full diploma program and is run on a space-available basis for individual courses.

Columbia College of Chicago

Address *600 S Michigan*
Chicago *IL*
60605 *USA*

Phone *312-663-1600* **FAX**

Director *Douglas Jones* **Admission Contact** *Douglas Jones*

Program Founded *1989* **School Type** *University*

Program Offered *Sound Technology*

Degrees Offered *Bachelor of Arts*

Program Length *4 years*

Estimated Tuition *$24,000*

Main Emphasis *Audio Engineering* **Program is** *Semi-technical*

Accreditations *North Central*

Number of Studios **Is school non-profit?** *No*

Types of Studios
◉ *Acoustic Research* ◉ *Electronic Music* ◉ *Film/Foley* ◉ *Radio*
◉ *Audio Research* ◉ *Music Recording* ○ *Television* ◉ *Video*

Types of Recording
◉ *Analog Multitrack* ◉ *MIDI Sequencing* ◉ *Video*
○ *Digital Multitrack* ○ *DAW* ○ *Film*

Other Resources
◉ *Professional Studios* ○ *Television Stations*
◉ *Radio Stations* ◉ *Theater Tech Dept*

Class Size Lecture *18* **Class Size Lab** *12*

Assistance
○ *Housing* ◉ *Scholarships* ◉ *Internships*
◉ *Financial Aid* ◉ *Work-study* ◉ *Job Placement*

Admission Policy *Open* **Language** *English*

Prerequisites ○ *High School Diploma* ○ *SAT-ACT* ○ *Music Audition*

Industry Affiliations
○ *AES* ○ *ASA* ○ *NACB* ○ *NARAS* ○ *SMPTE* ○ *NAMBI*
○ *APRS* ○ *NAB* ○ *NAMM* ○ *SBE* ○ *SPARS* ○ *MEIEA*

Fulltime Faculty *2* **Parttime Faculty** *26*

Faculty Awards

Program Awards

Student Awards

Research areas

Columbia College of Chicago

Columbia College of Chicago offers a Sound Technology major with three optional concentrations: Recording, Acoustics/Sound Contracting, and Sound for Pictures. The goal of the Sound Technology major is to educate people who want to work with audio in a variety of fields. The school's vision of the industry is broader than most programs. They see good career opportunities for students in live concert sound, AV production, sound system design, acoustics, pro-audio sales, film sound, television sound, and more. The Sound Major lays a foundation that can be used in any of these disciplines and goes on to explore specifics.

The Department of Radio/Sound offers more than 20 classes pertaining to professional audio. They range from introductory classes to advanced classes such as Time Delay Spectrometry, Computer-Aided Sound System Design and Modeling, and Advanced Digital Sound Recording. In addition to the offerings from the Department of Radio/Sound, there are 90 additional credit hours in other departments in the College that deal with some aspect of sound. The College also offers programs in music business and media management.

The Recording Concentration required courses include: Sound Engineering, Acoustics for Microphones, Audio Processing, Audio Technologies, Jingle Production, and Digital Sound. The Acoustics/Sound Contracting Concentration requires: Acoustics of Microphones, Acoustics II, Advanced Acoustic Design, Audio Equipment Overview, Sound System Design, and Time Delay Spectrometry. The Sound for Pictures Concentration includes: Audio Processing, Audio for the Visual Medium I & II, Digital Sound, and Audio Technologies.

All the faculty in the Sound major are professionals in their various fields. In addition to the commitment to their own careers in audio, they share a commitment to providing a well-rounded education. The major is directed by Douglas Jones, president of Electro Acoustic Systems, Inc.. He is an authority on studio and audio system design, and developed the LEDR™ test for stereo imagery while part of a research team at Northwestern University. Other faculty members include several active recording engineers, acoustic consultants, and system designers.

Conservatory of Recording Arts & Sciences

Address *1100 East Missouri, Ste 400*
Phoenix *AZ*
85014 *USA*

Phone *602-265-6383* **FAX** *602-230-7235*

Director *Kirk Hamm* **Admission Contact** *Kirk Hamm*

Program Founded *1980* **School Type** *Trade School*

Program Offered *Master Recording*

Degrees Offered *Diploma*

Program Length *15 weeks plus internship*

Estimated Tuition *$6,065*

Main Emphasis *Audio Engineering* **Program is** *Technical*

Accreditations *CCA/ACTTS*

Number of Studios *5* **Is school non-profit?** *No*

Types of Studios ○ *Acoustic Research* ◉ *Electronic Music* ○ *Film/Foley* ○ *Radio*
○ *Audio Research* ◉ *Music Recording* ○ *Television* ○ *Video*

Types of Recording ◉ *Analog Multitrack* ◉ *MIDI Sequencing* ◉ *Video*
◉ *Digital Multitrack* ○ *DAW* ○ *Film*

Other Resources ◉ *Professional Studios* ○ *Television Stations*
○ *Radio Stations* ○ *Theater Tech Dept*

Class Size Lecture *8* **Class Size Lab** *8*

Assistance ◉ *Housing* ○ *Scholarships* ◉ *Internships*
◉ *Financial Aid* ○ *Work-study* ◉ *Job Placement*

Admission Policy *Selective* **Language** *English*

Prerequisites ◉ *High School Diploma* ○ *SAT-ACT* ○ *Music Audition*
Two Letters of Reference

Industry Affiliations ◉ *AES* ○ *ASA* ○ *NACB* ○ *NARAS* ○ *SMPTE* ○ *NAMBI*
○ *APRS* ○ *NAB* ○ *NAMM* ○ *SBE* ◉ *SPARS* ○ *MEIEA*

Fulltime Faculty *2* **Parttime Faculty** *4*

Faculty Awards

Program Awards

Student Awards

Research areas

Conservatory of Recording Arts & Sciences

The Conservatory of Recording Arts & Sciences offers the Master Recording Program, a five course curriculum of 600 hours of instruction and internship experience. The school's objective is to train students for entry-level positions in a variety of areas in the music and recording industries.

The courses in the program include: Audio Recording & Production, Music Business, MIDI/Computer/Electronic Music Recording, Sound Reinforcement, and Troubleshooting/Maintenance. Students may register for individual courses, but Master Recording Program students have priority.

A 280-hour internship is the final phase of the Master Recording Program and is taken immediately following completion of coursework. Internships are in a studio or related facility. The school cannot guarantee a specific internship, but does attempt to work with students needs and wishes.

Dartmouth College- Music Department

Address	*6147 Hopkins Center* *Hanover* *NH* *03755* *USA*
Phone	*603-646-3974* **FAX** *604-646-2551*
Director	*Jon Appleton* **Admission Contact** *Suki Sodhi*
Program Founded	*1989* **School Type** *University*
Program Offered	*Electro-Acoustic Music*
Degrees Offered	*Master of Arts*
Program Length	*2 years*
Estimated Tuition	*None*
Main Emphasis	*Electronic Music* **Program is** *Technical*
Accreditations	*NASM*
Number of Studios	*4* **Is school non-profit?** *Yes*
Types of Studios	● *Acoustic Research* ● *Electronic Music* ○ *Film/Foley* ○ *Radio* ○ *Audio Research* ○ *Music Recording* ○ *Television* ○ *Video*
Types of Recording	○ *Analog Multitrack* ● *MIDI Sequencing* ○ *Video* ● *Digital Multitrack* ○ *DAW* ○ *Film*
Other Resources	● *Professional Studios* ○ *Television Stations* ○ *Radio Stations* ○ *Theater Tech Dept*
Class Size Lecture	*3* **Class Size Lab** *3*
Assistance	● *Housing* ● *Scholarships* ○ *Internships* ● *Financial Aid* ○ *Work-study* ○ *Job Placement*
Admission Policy	*Highly selective* **Language** *English*
Prerequisites	○ *High School Diploma* ○ *SAT-ACT* ○ *Music Audition* *GRE*
Industry Affiliations	○ *AES* ○ *ASA* ○ *NACB* ○ *NARAS* ○ *SMPTE* ○ *NAMBI* ○ *APRS* ○ *NAB* ○ *NAMM* ○ *SBE* ○ *SPARS* ○ *MEIEA*
Fulltime Faculty	*3* **Parttime Faculty** *6*
Faculty Awards	
Program Awards	
Student Awards	
Research areas	*Music, Engineering, and Computer Science*

Dartmouth College- Music Department

Dartmouth College's Department of Music offers a unique graduate program in Electro-Acoustic Music. The two-year study is aimed at students interested in combining music, engineering, and computer science classes into a comprehensive program emphasizing electro-acoustic music. Applicants to the Master of Arts program must have an bachelor's degree in either music, engineering, or computer science to be considered for admission. The program only admits three students a year. Those selected receive full tuition fellowships as well as a generous stipend.

The requirements of the program include six terms of residency in the program, musical ability as well as knowledge of music theory and history, several seminars in music and technology, several courses in psychology, acoustics, computer science, engineering sciences, and electives. The program also requires directed research, resulting in a graduate thesis. Students benefit from the Bregman Electronic Music Studio and related facilities.

The faculty members involved in the program are from a variety of departments, including Music, Psychology, Physics, Mathematics and Computer Science, and the Thayer School of Engineering.

DePaul University School of Music

Address *804 West Belden Avenue*
Chicago *IL*
60614 *USA*

Phone *312-362-6844* **FAX** *312-362-8215*

Director *Dean E. Kocher* **Admission Contact** *Robert Shamo*

Program Founded *1987* **School Type** *University*

Program Offered *Sound Recording Technology*

Degrees Offered *Bachelor of Science in Music: Elective Studies in Electrical Engineering*

Program Length *4 years*

Estimated Tuition *$45,000*

Main Emphasis *Audio Engineering* **Program is** *Technical*

Accreditations

Number of Studios *7* **Is school non-profit?** *Yes*

Types of Studios
○ *Acoustic Research* ○ *Electronic Music* ○ *Film/Foley* ○ *Radio*
○ *Audio Research* ◉ *Music Recording* ○ *Television* ○ *Video*

Types of Recording
◉ *Analog Multitrack* ◉ *MIDI Sequencing* ○ *Video*
◉ *Digital Multitrack* ○ *DAW* ○ *Film*

Other Resources
◉ *Professional Studios* ○ *Television Stations*
○ *Radio Stations* ○ *Theater Tech Dept*

Class Size Lecture *12* **Class Size Lab** *12*

Assistance
◉ *Housing* ◉ *Scholarships* ◉ *Internships*
◉ *Financial Aid* ◉ *Work-study* ◉ *Job Placement*

Admission Policy *Selective* **Language** *English*

Prerequisites ◉ *High School Diploma* ◉ *SAT-ACT* ◉ *Music Audition*

Industry Affiliations
○ *AES* ○ *ASA* ○ *NACB* ◉ *NARAS* ○ *SMPTE* ○ *NAMBI*
○ *APRS* ○ *NAB* ○ *NAMM* ○ *SBE* ○ *SPARS* ○ *MEIEA*

Fulltime Faculty **Parttime Faculty** *4*

Faculty Awards

Program Awards

Student Awards

Research areas

DePaul University School of Music

Offered in association with Steeterville Studios, DePaul's professional degree program prepares students for careers as recording engineers and sound technicians. The program includes requirements in liberal studies, core requirements in music, sound recording technology requirements, and electives in math, physics, electrical engineering, and computer science.

The School of Music at DePaul provides recording students with a variety of music to record. Students gain practical benefits of continuous experience in recording projects ranging fro chamber ensembles, jazz bands and orchestras to rock bands and computer-based composition. The Sound Recording Technology program balances the theoretical and practical sides of professional training.

The program benefits from Studio DePaul, the school's on-campus recording facility. The studio is tied to the school's recital hall, music room, and CAI lab. Established in 1992, the studio serves as an adjunct to the primary downtown facility of Streeterville Studios, as the training ground for DePaul's Sound Recording Technology majors.

Elizabeth City State University

Address *Campus Box 809*
Elizabeth City *NC*
27909 *USA*

Phone *919-335-3377* **FAX**

Director *Scott Fredrickson* **Admission Contact** *Scott Fredrickson*

Program Founded *1980* **School Type** *University*

Program Offered *Music Engineering & Technology/Music Business Administration*

Degrees Offered *Bachelor of Science in Music Industry Studies*

Program Length *4 years*

Estimated Tuition *Inquire*

Main Emphasis *Audio Engineering* **Program is** *Technical*

Accreditations

Number of Studios *2* **Is school non-profit?** *Yes*

Types of Studios
- ○ *Acoustic Research* ○ *Electronic Music* ○ *Film/Foley* ○ *Radio*
- ○ *Audio Research* ◉ *Music Recording* ○ *Television* ○ *Video*

Types of Recording
- ○ *Analog Multitrack* ◉ *MIDI Sequencing* ○ *Video*
- ○ *Digital Multitrack* ○ *DAW* ○ *Film*

Other Resources
- ◉ *Professional Studios* ○ *Television Stations*
- ◉ *Radio Stations* ○ *Theater Tech Dept*

Class Size Lecture *6* **Class Size Lab** *6*

Assistance
- ◉ *Housing* ◉ *Scholarships* ◉ *Internships*
- ◉ *Financial Aid* ◉ *Work-study* ◉ *Job Placement*

Admission Policy *Selective* **Language** *English*

Prerequisites ◉ *High School Diploma* ◉ *SAT-ACT* ○ *Music Audition*

Industry Affiliations
- ◉ *AES* ○ *ASA* ○ *NACB* ◉ *NARAS* ○ *SMPTE* ◉ *NAMBI*
- ○ *APRS* ○ *NAB* ◉ *NAMM* ○ *SBE* ○ *SPARS* ◉ *MEIEA*

Fulltime Faculty *2* **Parttime Faculty**

Faculty Awards

Program Awards

Student Awards

Research areas

Elizabeth City State University

Elizabeth City State University has developed an alternative to the strictly traditional music degree program with its Bachelor of Science in Music Industry Studies. The program combines a traditional music curriculum with industry related courses and experiences to prepare well-rounded graduates who are knowledgeable in all aspects of the music industry.

Within the Music Industry Studies program, Elizabeth City State University offers concentrations in Music Engineering & Technology or Music Business Administration. The Music Engineering & Technology concentration is based on state-of-the-art 24-track recording and MIDI/electronic music production. The curriculum incorporates studies in studio recording, production, digital audio, MIDI/electronic music, computer applications, acoustics, and studio design. Students develop the technical skills and creativity necessary to be competitive and successful in the recording industry. The programs studios enable students to produce, record, and market their projects as a major component of instruction.

The Music Business Administration concentration focuses on music business, management, marketing, sales, publishing, record production, retailing, promotion, and live concert production. The Music Industry Studies program plans to set up several student-operated enterprises. These businesses will include a record label, music publishing company, and production company. The activities from these business will provide real-world music industry experience for the program's students.

Currently under development is an Electronic Music Production Studio/Media Resource Center. The Media Resource Center will provide students will convenient access to music industry related texts, periodicals, professional journals, audio recordings, software, and video tapes. The Production Studio will provide the Music Department with capabilities for MIDI/electronic music composition and production as well as computer-assisted music theory and aural skills training. In addition, the Media Resources Center will be the administrative headquarters for the student-operated record label and music publishing and production companies.

The faculty in charge of the development and instruction of these programs are specialists in their fields. Dr. Scott Fredrickson, Director of Music Industry Studies, has extensive experience within the music industry and holds degrees in music education, business administration, jazz, and music business administration. Barry Hill, Director of Music Engineering & Technology, is experienced in recording engineering, studio management, MIDI/electronic music, and live sound reinforcement, and has degrees in music and recording arts.

Elmhurst College

Address *190 Prospect*
Elmhurst *IL*
60126 *USA*

Phone *708-617-3515* **FAX** *708-617-3245*

Director *Tim Hays* **Admission Contact** *Doug Beach*

Program Founded *1972* **School Type** *University*

Program Offered *Music Business with Audio Engineering Emphasis*

Degrees Offered *Bachelor of Science, Bachelor of Music*

Program Length *4 years*

Estimated Tuition *$37,000*

Main Emphasis *Music Business* **Program is** *Non-technical*

Accreditations *North Central*

Number of Studios *3* **Is school non-profit?** *Yes*

Types of Studios
○ *Acoustic Research* ◉ *Electronic Music* ○ *Film/Foley* ○ *Radio*
○ *Audio Research* ◉ *Music Recording* ○ *Television* ◉ *Video*

Types of Recording
◉ *Analog Multitrack* ◉ *MIDI Sequencing* ○ *Video*
○ *Digital Multitrack* ○ *DAW* ○ *Film*

Other Resources
◉ *Professional Studios* ○ *Television Stations*
◉ *Radio Stations* ◉ *Theater Tech Dept*

Class Size Lecture *15* **Class Size Lab** *6*

Assistance
◉ *Housing* ◉ *Scholarships* ◉ *Internships*
◉ *Financial Aid* ◉ *Work-study* ◉ *Job Placement*

Admission Policy *Selective* **Language** *English*

Prerequisites ◉ *High School Diploma* ◉ *SAT-ACT* ◉ *Music Audition*

Industry Affiliations
◉ *AES* ○ *ASA* ○ *NACB* ◉ *NARAS* ○ *SMPTE* ◉ *NAMBI*
○ *APRS* ○ *NAB* ◉ *NAMM* ○ *SBE* ○ *SPARS* ◉ *MEIEA*

Fulltime Faculty *5* **Parttime Faculty** *42*

Faculty Awards

Program Awards

Student Awards

Research areas

Elmhurst College

Elmhurst College offers either Bachelor of Science or Bachelor of Music degrees in Music Business. Students interested in audio engineering and music recording can structure their programs to emphasis these areas. The curriculums for both degree programs blend courses from music, music business, and business with a core of liberal arts courses. The Bachelor of Music program includes additional courses in music, while the Bachelor of Science has additional courses in business and mathematics.

The Music Business program is designed to accommodate the diversity of opportunities in the music industry by allowing students to pursue their own interests. Elmhurst College's location in the Metropolitan Chicago Area enables students to take advantage of internship possibilities in the local music industry and to benefit from guest speakers who are professionally active in the area. The program also organizes field trips to other music business centers, such as New York, Los Angeles, and Germany.

Recording facilities include a multitrack studio with digital mastering capabilities housed in the school's Computer Science and Technology Center. The program also utilizes a MIDI lab for electronic music production and synthesis. The recording facility has been ranked as the best college studio in Illinois.

Film & Television Institute of India

Address	*Law College Road*		
	Pune	*Maharashtra*	
	411 004	*India*	
Phone	*331113*	**FAX**	-
Director	*Shri Satish Kumar*	**Admission Contact**	*Registrar*
Program Founded	*1967*	**School Type**	*Trade School*
Program Offered	*Sound Recording for Film & Video*		
Degrees Offered	*Diploma-in-Cinema*		
Program Length	*3 years*		
Estimated Tuition	*$120 for residents*		
Main Emphasis	*Film/Video Sound*	**Program is**	*Semi-technical*
Accreditations			
Number of Studios		**Is school non-profit?**	*Yes*

Types of Studios
- ○ *Acoustic Research* ○ *Electronic Music* ◉ *Film/Foley* ○ *Radio*
- ○ *Audio Research* ○ *Music Recording* ○ *Television* ○ *Video*

Types of Recording
- ◉ *Analog Multitrack* ○ *MIDI Sequencing* ○ *Video*
- ○ *Digital Multitrack* ○ *DAW* ◉ *Film*

Other Resources
- ○ *Professional Studios* ○ *Television Stations*
- ○ *Radio Stations* ○ *Theater Tech Dept*

Government Financed

Class Size Lecture	*varies*	**Class Size Lab**	*varies*

Assistance
- ○ *Housing* ◉ *Scholarships* ○ *Internships*
- ○ *Financial Aid* ○ *Work-study* ○ *Job Placement*

Admission Policy	*Highly selective*	**Language**	*English*

Prerequisites
- ○ *High School Diploma* ○ *SAT-ACT* ○ *Music Audition*

Degree Science

Industry Affiliations
- ○ *AES* ○ *ASA* ○ *NACB* ○ *NARAS* ○ *SMPTE* ○ *NAMBI*
- ○ *APRS* ○ *NAB* ○ *NAMM* ○ *SBE* ○ *SPARS* ○ *MEIEA*

Fulltime Faculty	*6*	**Parttime Faculty**	*10*
Faculty Awards			
Program Awards			
Student Awards			
Research areas			

Film & Television Institute of India

The Film & Television Institute of India offers a three-year program in Sound Recording and Electronics. The program consists of six 20-week semesters. Admission is limited to ten students per course, eight open and two reserved for Afro-Asian students. There are four disciplines: Film Direction, Motion Picture Photography, Sound Recording and Engineering, and Film Editing. The minimum qualification for admission is graduate for Film Direction and Film Editing and Science graduate for Motion Picture Photography and Sound Recording.

In the program's first year, all students are required to take two semesters of common courses. These are designed to familiarize students with all aspects of film making. At the end of the first year, students split into specialized studies in their chosen field.

During the second year, intensive training is imparted in all aspects of film sound recording and related electronics. Emphasis is balanced between recording and electronics. Students are trained for dubbing, mixing, and music recording. In electronics, students are trained in basic principles of analog and digital electronics. Qualifying students move on to the third year of studies.

In the final year, all students come together and work through a series of film exercises. During this period, students refine their area of specialization. The final exercise is a 30 minute black and white 35 mm diploma film. Students are also required to complete an electronic project. Since students are trained in both electronics and recording, they are able to work either as a recordist or a service engineer in the film and television industry.

Finger Lakes Community College

Address	*4355 Lake Shore Drive* *Canandaigua* *14424-8395*	*NY* *USA*	
Phone	*716-394-3500*	**FAX**	*716-394-5005*
Director	*Frank Verget*	**Admission Contact**	*John Meuser*
Program Founded	*1979*	**School Type**	*Community College*
Program Offered	*Music Recording, Broadcasting Communications*		
Degrees Offered	*Associate in Arts Liberal Arts and Sciences*		
Program Length	*2 years*		
Estimated Tuition	*$2,800*		
Main Emphasis	*Music Recording*	**Program is**	*Semi-technical*
Accreditations	*Middle States*		
Number of Studios	*2*	**Is school non-profit?**	*Yes*
Types of Studios	○ *Acoustic Research* ○ *Electronic Music* ○ *Film/Foley* ○ *Radio* ○ *Audio Research* ◉ *Music Recording* ○ *Television* ○ *Video*		
Types of Recording	◉ *Analog Multitrack* ○ *MIDI Sequencing* ◉ *Video* ○ *Digital Multitrack* ○ *DAW* ○ *Film*		
Other Resources	◉ *Professional Studios* ○ *Television Stations* ○ *Radio Stations* ○ *Theater Tech Dept*		
Class Size Lecture	*15*	**Class Size Lab**	*4*
Assistance	◉ *Housing* ◉ *Scholarships* ◉ *Internships* ◉ *Financial Aid* ◉ *Work-study* ◉ *Job Placement*		
Admission Policy	*Open*	**Language**	*English*
Prerequisites	◉ *High School Diploma* ○ *SAT-ACT* ○ *Music Audition*		
Industry Affiliations	○ *AES* ○ *ASA* ○ *NACB* ○ *NARAS* ○ *SMPTE* ○ *NAMBI* ○ *APRS* ○ *NAB* ○ *NAMM* ○ *SBE* ○ *SPARS* ○ *MEIEA*		
Fulltime Faculty	*3*	**Parttime Faculty**	*8*
Faculty Awards			
Program Awards			
Student Awards			
Research areas			

Finger Lakes Community College

Finger Lakes Community College offers two associate degree programs of interest: Broadcasting Communications and Music Recording. The two-year programs are based in liberal arts studies to enable graduates to continue their educations and pursue baccalaureate degrees at four-year institutions. Emphases are provided to permit students to focus their educational experience to fit their interests and career goals.

The Broadcasting Communications emphasis offers hands-on experience in camera operation, directing, writing, graphics production, performance, and videotape editing techniques. The curriculum also provides a background in business, writing, and theater. Students work in the College's local television station and professionally-equipped studio. Students also gain experience by working on location. Internships are available on campus and at area television stations and video production houses.

The Music Recording emphasis is designed for students interested in sound recording and related areas of the music business. Students who complete the degree may transfer to four-year recording programs or immediately enter a career in the music industry. Students with a background in music are encouraged to take music theory courses, and mathematics and physics courses are recommended electives.

Finger Lakes Community College has a variety of organizations to accommodate the diverse interest of students. The Communications Club promotes an interest in broadcasting and public relations and participates in conferences and field trips to television stations and video production houses. The school also has an active Radio Club that provides entertainment and publicizes campus activities.

Five Towns College

Address *305 North Service Road*
Dix Hills *NY*
11746-6055 *USA*

Phone *516-424-7000* **FAX** *516-424-7006*

Director *Martin Cohen* **Admission Contact** *Jennifer Roemer*

Program Founded *1972* **School Type** *University*

Program Offered *Audio Recording Technology*

Degrees Offered *Associate in Applied Science & Bachelor of Music*

Program Length *2 years, 4 years*

Estimated Tuition *$6,150 per year*

Main Emphasis *Music Recording* **Program is** *Semi-technical*

Accreditations *Middle State Assoc., NYS Board of Regents*

Number of Studios *4* **Is school non-profit?** *No*

Types of Studios
◉ *Acoustic Research* ◉ *Electronic Music* ○ *Film/Foley* ○ *Radio*
◉ *Audio Research* ◉ *Music Recording* ○ *Television* ◉ *Video*

Types of Recording
◉ *Analog Multitrack* ◉ *MIDI Sequencing* ◉ *Video*
◉ *Digital Multitrack* ○ *DAW* ○ *Film*

Other Resources
◉ *Professional Studios* ◉ *Television Stations*
○ *Radio Stations* ○ *Theater Tech Dept*

Class Size Lecture *40* **Class Size Lab** *15*

Assistance
◉ *Housing* ◉ *Scholarships* ◉ *Internships*
◉ *Financial Aid* ◉ *Work-study* ◉ *Job Placement*

Admission Policy *Selective* **Language** *English*

Prerequisites ◉ *High School Diploma* ○ *SAT-ACT* ◉ *Music Audition*

Industry Affiliations
◉ *AES* ○ *ASA* ○ *NACB* ◉ *NARAS* ◉ *SMPTE* ○ *NAMBI*
○ *APRS* ◉ *NAB* ◉ *NAMM* ○ *SBE* ◉ *SPARS* ○ *MEIEA*

Fulltime Faculty *33* **Parttime Faculty** *34*

Faculty Awards

Program Awards

Student Awards

Research areas

Five Towns College

The Audio Recording Technology concentration at Five Towns College is designed to provide students with the tools needed to begin careers as professional music engineers and producers of soundtracks for film and video productions. Students receive intensive instruction in audio recording, a comprehensive music education, and the opportunity for supervised internships.

Students study the theory of sound, recording electronics, engineering procedures, music production techniques, and audio/video post-production in a sequence of courses designed to develop practical and technical skills. The college's state-of-the-art, computer-automated multitrack recording studio, editing suite, and individual MIDI workstations provide students with a highly focused learning environment for assigned and extracurricular recording projects.

Five Towns College's internship and job placement experiences provide students with professional training outside the classroom. The college has placed students in almost every major recording facility in the New York City metropolitan region. Internships, combined with electives courses in music business, contracts, law, promotion, and broadcasting, give students a competitive edge in the audio recording marketplace.

The school offers both an Associate in Applied Sciences and a Bachelor of Music degree in this area. Other related programs of interest at Five Towns College are: Music Business, Video Music, Jazz/Commercial Music, and Music Instrument Technology.

Full Sail Center for the Recording Arts

Address	*3300 University Blvd., Suite 160* *Winter Park* *FL* *32792* *USA*		
Phone	*407-679-6333*	**FAX**	*407-678-0070*
Director	*Gerry Platt*	**Admission Contact**	*Gerry Platt*
Program Founded	*1979*	**School Type**	*Trade School*
Program Offered	*Recording Arts, Video & Film Production*		
Degrees Offered	*Specialized Associate Degree*		
Program Length	*54 weeks*		
Estimated Tuition	*$17,500*		
Main Emphasis	*Audio Engineering*	**Program is**	*Technical*
Accreditations	*CCA*		
Number of Studios	*11*	**Is school non-profit?**	*No*
Types of Studios	○ *Acoustic Research* ◉ *Electronic Music* ○ *Film/Foley* ○ *Radio* ○ *Audio Research* ◉ *Music Recording* ○ *Television* ○ *Video*		
Types of Recording	◉ *Analog Multitrack* ◉ *MIDI Sequencing* ◉ *Video* ◉ *Digital Multitrack* ◉ *DAW* ◉ *Film*		
Other Resources	◉ *Professional Studios* ○ *Television Stations* ○ *Radio Stations* ○ *Theater Tech Dept*		
Class Size Lecture	*45*	**Class Size Lab**	*6*
Assistance	◉ *Housing* ○ *Scholarships* ◉ *Internships* ◉ *Financial Aid* ○ *Work-study* ◉ *Job Placement*		
Admission Policy	*Open*	**Language**	*English*
Prerequisites	◉ *High School Diploma* ○ *SAT-ACT* ○ *Music Audition*		
Industry Affiliations	◉ *AES* ○ *ASA* ○ *NACB* ◉ *NARAS* ◉ *SMPTE* ○ *NAMBI* ○ *APRS* ◉ *NAB* ◉ *NAMM* ○ *SBE* ◉ *SPARS* ◉ *MEIEA*		
Fulltime Faculty	*52*	**Parttime Faculty**	*28*
Faculty Awards	*Crystal Reel, TEC, Gold & Platinum Records, Addy's & Emmy's*		
Program Awards	*TEC Award- 3 years*		
Student Awards	*Platinum Dream Award, Grammy Nominations, Emmy's, Gold & Platinum*		
Research areas	*DAW and outboard gear beta site*		

Full Sail Center for the Recording Arts

Full Sail Center for the Recording Arts offers training for career in the entertainment industry centered around video, film, and audio production. The school is located in Winter Park, Florida, a suburb of Orlando, one of the United States' fastest growing markets for video and film production. Full Sail's philosophy of education is based on three basic concepts: 1) students must be trained in a real-world, working media complex that utilizes the latest technology in film, video, and audio equipment; 2) Full Sail students must be taught by industry professionals still active in their respective fields of expertise; 3) the school must be centered in the heart of the industry for which it trains students. From the school's inception in 1978, Full Sail has followed these three principles.

The educational media complex consists of ten state-of-the-art media studios that are part of the 70,000 square foot campus. The studios contain the latest technology that the industry has to offer and benefit from Full Sail being a training center and beta test site for several specialized audio products from major manufactures. The school also features a complete mobile recording unit. Hands-on training is emphasized at Full Sail with approximately 40% of the educational hours dedicated to labs in the studios. Students learn operating techniques, industry protocol, and equipment maintenance. Students also receive an introduction to numerous career fields, including the requirements of the student's chosen career, income potential, lifestyle, and the current state of the job market.

A low student to faculty ratio of 6 to 1 is maintained for all labs, allowing students to fulling interact with instructors, course directors, and special entertainment guests. Full Sail offers financial aid to all who qualify, and features a job placement department utilizing the services of four, fulltime employees whose sole mission is to aid graduates with acquiring internships and employment.

Georgia State University School of Music

Address	*University Plaza* *Atlanta* *30303*	*GA* *USA*	
Phone	*404-651-3676*	**FAX**	*404-525-4545*
Director	*James Oliverio*	**Admission Contact**	*Dr. Harding*
Program Founded	*1978*	**School Type**	*University*
Program Offered	*Sound Recording Technology*		
Degrees Offered	*Bachelor of Music*		
Program Length	*4 years*		
Estimated Tuition	*$8,000*		
Main Emphasis	*Music Recording*	**Program is**	*Semi-technical*
Accreditations	*NASM*		
Number of Studios	*2*	**Is school non-profit?**	*Yes*
Types of Studios	○ *Acoustic Research* ◉ *Electronic Music* ○ *Film/Foley* ○ *Radio* ○ *Audio Research* ◉ *Music Recording* ○ *Television* ○ *Video*		
Types of Recording	◉ *Analog Multitrack* ◉ *MIDI Sequencing* ◉ *Video* ○ *Digital Multitrack* ◉ *DAW* ○ *Film*		
Other Resources	◉ *Professional Studios* ○ *Television Stations* ○ *Radio Stations* ○ *Theater Tech Dept*		
Class Size Lecture	*12*	**Class Size Lab**	*6*
Assistance	◉ *Housing* ◉ *Scholarships* ◉ *Internships* ◉ *Financial Aid* ◉ *Work-study* ◉ *Job Placement*		
Admission Policy	*Selective*	**Language**	*English*
Prerequisites	◉ *High School Diploma* ◉ *SAT-ACT* ◉ *Music Audition*		
Industry Affiliations	○ *AES* ○ *ASA* ○ *NACB* ○ *NARAS* ○ *SMPTE* ○ *NAMBI* ○ *APRS* ○ *NAB* ○ *NAMM* ○ *SBE* ◉ *SPARS* ○ *MEIEA*		
Fulltime Faculty	*3*	**Parttime Faculty**	*5*
Faculty Awards			
Program Awards			
Student Awards			
Research areas	*Music for Film/Video- Audio Post-Production*		

Georgia State University School of Music

The Sound Recording Technology program at the Georgia State University School of Music is an emphasis within the school's Music Industry program. The school also offers an emphasis in Music Business Management. The goal of the Recording track is to give students both the theoretical background and the practical foundation required to pursue a career as either a music producer, engineer or both.

The program's curriculum is balanced between a core of music and liberal arts and sciences courses and music business and recording classes. In 1994, the School of Music is relocating to a new facility, and the scope of the Sound Recording Technology program will be enhanced by facilities for live recording, including a 1,200 seat concert hall, several recording labs, and related production facilities. The program also benefits from the performing talents of the School of Music's students. The intent of the program is to allow students to acquire experience with a variety of recording and producing, from classical ensembles to post-production for film and video.

The internship program at Georgia State University is the foundation of the Music Industry program. Students can take many different positions to help them determine their career goals. Unlike most schools, students can begin taking internships in their freshman year after completing the English composition sequence and the Introduction of the Music Industry class. Interns benefit from Atlanta's growing audio and post-production market.

James Oliverio, Director of Music Technology, is an Emmy Award-winning composer and music producer. He is experienced with more than 300 film and television soundtracks, as well as feature film scores and major works for the concert hall.

Global Media Institute of Arts & Sciences

Address *6325 N Invergordon*
Paradise Valley *AZ*
85253 *USA*

Phone *602-948-5883* **FAX** *602-948-7863*

Director *Dave Cornelius* **Admission Contact** *Steve Brittle*

Program Founded *1992* **School Type** *Trade School*

Program Offered *Recording Engineering*

Degrees Offered *Diploma and Certificate*

Program Length *6 months*

Estimated Tuition *$6,000*

Main Emphasis *Audio Engineering* **Program is** *Semi-technical*

Accreditations

Number of Studios *3* **Is school non-profit?** *No*

Types of Studios ○ *Acoustic Research* ◉ *Electronic Music* ○ *Film/Foley* ○ *Radio*
○ *Audio Research* ◉ *Music Recording* ○ *Television* ○ *Video*

Types of Recording ◉ *Analog Multitrack* ◉ *MIDI Sequencing* ◉ *Video*
○ *Digital Multitrack* ◉ *DAW* ○ *Film*

Other Resources ◉ *Professional Studios* ○ *Television Stations*
○ *Radio Stations* ○ *Theater Tech Dept*

Class Size Lecture *6-10* **Class Size Lab** *6-10*

Assistance ◉ *Housing* ○ *Scholarships* ◉ *Internships*
◉ *Financial Aid* ○ *Work-study* ◉ *Job Placement*

Admission Policy *Open* **Language** *English*

Prerequisites ◉ *High School Diploma* ○ *SAT-ACT* ○ *Music Audition*

Industry Affiliations ○ *AES* ○ *ASA* ○ *NACB* ○ *NARAS* ○ *SMPTE* ○ *NAMBI*
○ *APRS* ○ *NAB* ○ *NAMM* ○ *SBE* ○ *SPARS* ○ *MEIEA*

Fulltime Faculty *2* **Parttime Faculty** *2*

Faculty Awards

Program Awards

Student Awards

Research areas *DAW, Interactive TV, Fiber Optic Delivery Systems, Distance Learning*

Global Media Institute of Arts & Sciences

GMIAS offers an Audio Recording Engineering program comprised of a seven-course curriculum divided into 600 hours of instruction. The objective of the program is to train students for entry level positions in a variety of areas in the music and recording industries. The program consists of the following courses: Intro to Audio Recording/Production, Audio Recording/Production Practicum, MIDI/Computer/Electronic Music Theory and Applications, Troubleshooting and Maintenance, Business of Music, Recording Lab/Individual Projects, and Internship.

The 280 internship is the final phase of the program and is to be taken following completion of coursework. Internships should begin within 30 days after completing classes and lab work, and should be based upon a 40-hour week. All interns are supervised and evaluated by their immediate facility advisor and are observed and counseled during site visits by the school's Director of Education or his designated case representative.

Tho program is 16 weeks long, 20 hours per week, plus a seven-week internship. Classes are scheduled in four-hour blocks, Monday through Friday, with morning and evening sessions. GMIAS is part of Pantheon Studios, Inc., an active audio, video, and film production company.

Hennepin Technical College

Address	*9200 Flying Cloud Drive*		
	Edin Prairie	*MN*	
	55347	*USA*	
Phone	*612-944-2222*	**FAX**	*612-550-3147*
Director	*Tom Winger*	**Admission Contact**	*Tim Standorfer*
Program Founded	*1989*	**School Type**	*Trade School*
Program Offered	*Audio Recording Specialist*		
Degrees Offered	*Diploma*		
Program Length	*2 years*		
Estimated Tuition	*$4,000*		
Main Emphasis	*Audio Engineering*	**Program is**	*Technical*
Accreditations			
Number of Studios	*6*	**Is school non-profit?**	*Yes*

Types of Studios
○ *Acoustic Research* ◉ *Electronic Music* ◉ *Film/Foley* ○ *Radio*
○ *Audio Research* ◉ *Music Recording* ○ *Television* ○ *Video*

Types of Recording
◉ *Analog Multitrack* ◉ *MIDI Sequencing* ○ *Video*
○ *Digital Multitrack* ◉ *DAW* ○ *Film*

Other Resources
◉ *Professional Studios* ◉ *Television Stations*
○ *Radio Stations* ○ *Theater Tech Dept*

Class Size Lecture	*25*	**Class Size Lab**	*varies*

Assistance
○ *Housing* ◉ *Scholarships* ◉ *Internships*
◉ *Financial Aid* ◉ *Work-study* ◉ *Job Placement*

Admission Policy	*Open*	**Language**	*English*

Prerequisites
○ *High School Diploma* ○ *SAT-ACT* ○ *Music Audition*
HTC Placement Test

Industry Affiliations
○ *AES* ○ *ASA* ○ *NACB* ○ *NARAS* ○ *SMPTE* ○ *NAMBI*
○ *APRS* ○ *NAB* ○ *NAMM* ○ *SBE* ◉ *SPARS* ○ *MEIEA*

Fulltime Faculty	*2*	**Parttime Faculty**	*6*

Faculty Awards

Program Awards

Student Awards

Research areas

Hennepin Technical College

Hennepin Technical College's Audio Recording Specialist program presents a well-rounded overview of the recording industry and prepares students for entry level positions in recording-related occupations. The rigorous six-quarter curriculum exposes students to a mixture of courses in recording techniques, acoustic design, electronics, and music fundamentals. The program's technical facilities expose students to the types of recoding hardware and software they will encounter after graduation.

Topics covered in the program include: multitrack recording, MIDI/SMPTE, audio for video and film, broadcast production, corporate/industrial production, basic electronics, and location audio. In addition to these, students may take related courses in television and video production, audio-visual media production, photography, computer graphics, and electronic publishing. Students are also offered the opportunity to intern at one of many professional media facilities in the Twin Cities metro area.

Hutchinson Vocational Technical Institute

Address *2 Century Avenue*
Hutchinson *MN*
55350 *USA*

Phone *612-587-3636* **FAX** *612-587-9091*

Director *David Igl* **Admission Contact** *David Igl*

Program Founded *1979* **School Type** *Community College*

Program Offered *Audio Technology*

Degrees Offered *Diploma and Associate in Applied Sciences*

Program Length *2 years*

Estimated Tuition *$4,000*

Main Emphasis *Maintenance* **Program is** *Technical*

Accreditations *North Central*

Number of Studios *2* **Is school non-profit?** *Yes*

Types of Studios
- ◉ *Acoustic Research*
- O *Electronic Music*
- O *Film/Foley*
- O *Radio*
- O *Audio Research*
- ◉ *Music Recording*
- O *Television*
- O *Video*

Types of Recording
- ◉ *Analog Multitrack*
- O *MIDI Sequencing*
- O *Video*
- O *Digital Multitrack*
- O *DAW*
- O *Film*

Other Resources
- O *Professional Studios*
- O *Television Stations*
- O *Radio Stations*
- O *Theater Tech Dept*

Class Size Lecture *10* **Class Size Lab** *12*

Assistance
- O *Housing*
- ◉ *Scholarships*
- ◉ *Internships*
- ◉ *Financial Aid*
- ◉ *Work-study*
- ◉ *Job Placement*

Admission Policy *Open* **Language** *English*

Prerequisites
- ◉ *High School Diploma*
- O *SAT-ACT*
- O *Music Audition*

Industry Affiliations
- ◉ *AES*
- O *ASA*
- O *NACB*
- O *NARAS*
- O *SMPTE*
- O *NAMBI*
- O *APRS*
- O *NAB*
- O *NAMM*
- O *SBE*
- ◉ *SPARS*
- O *MEIEA*

Fulltime Faculty *1* **Parttime Faculty** *2*

Faculty Awards

Program Awards

Student Awards

Research areas

Hutchinson Vocational Technical Institute

The Audio Technology Department at Hutchinson Technical College offers courses designed to give students knowledge and experience in both pro sound equipment applications and the electronic fundamentals which form the basis of operation, care, and maintenance of the gear. The program prepares students to install, operate, adjust, and diagnose major difficulties in audio equipment. Unlike other audio programs that concentrate predominately on recording, the Hutchinson program strives to give students a balanced curriculum with equal emphasis on technical aspects and a variety of applications, including sound reinforcement, acoustics, system installation and operation, and recording.

The Audio Technology program offers a challenging curriculum that is balanced between theory and operation of all types of audio equipment and the technical understanding of analog and digital electronics used to make the equipment function. Students planning to enroll should have or be willing to obtain a strong background in math balanced with a well-developed ear for music. A background in music, theater, math, physics, and electronics are all helpful. An interest in tinkering with and fixing electronic paraphernalia such as stereos, electric guitars, keyboards, amplifiers, and computers is very desirable. Many Audio Technology students are musicians, however, it is not required that you play an instrument to be a successful audio tech.

The 96-credit program can be completed in approximately two years, depending on the background of the individual. The instructional philosophy of HTC allows students to work at an accelerated rate, independent of other students in the program. Many students work quickly through areas they have previous work experience in and are able to spend more time in advanced areas. Group discussions and demonstrations are used in the courses to encourage students to seek additional information and clarification of course materials. The program tries to encourage students to get as much experience with the equipment as possible. Students are encouraged to use equipment above and beyond the basic course requirements.

The program's Internships are similar to the apprenticeship programs used by the electrician and medical professions. Students are expected to contribute to the profit motive of the internship employer and are often retained as permanent employees upon completion of the internship. Employers of the programs graduates support the internship concept as an excellent means of obtaining personnel with the job experience necessary for this industry.

Indiana University

Address *School of Music- Audio Department*
Bloomington *IN*
47405 *USA*

Phone *812-855-1087* **FAX** *812-855-4936*

Director *David Pickett* **Admission Contact** *David Pickett*

Program Founded *1981* **School Type** *University*

Program Offered *Audio Recording*

Degrees Offered *Associate of Science, Bachelor of Science*

Program Length *2 years, 4 years*

Estimated Tuition *$3,000-$8,600 per year*

Main Emphasis *Music Recording* **Program is** *Technical*

Accreditations *NASM*

Number of Studios *3* **Is school non-profit?** *Yes*

Types of Studios ○ *Acoustic Research* ○ *Electronic Music* ○ *Film/Foley* ○ *Radio*
○ *Audio Research* ◉ *Music Recording* ○ *Television* ○ *Video*

Types of Recording ◉ *Analog Multitrack* ○ *MIDI Sequencing* ○ *Video*
◉ *Digital Multitrack* ○ *DAW* ○ *Film*

Other Resources ○ *Professional Studios* ◉ *Television Stations*
◉ *Radio Stations* ◉ *Theater Tech Dept*

Class Size Lecture *15* **Class Size Lab** *8*

Assistance ◉ *Housing* ◉ *Scholarships* ◉ *Internships*
◉ *Financial Aid* ◉ *Work-study* ◉ *Job Placement*

Admission Policy *Highly selective* **Language** *English*

Prerequisites ◉ *High School Diploma* ◉ *SAT-ACT* ◉ *Music Audition*
Departmental Prerequisites

Industry Affiliations ○ *AES* ○ *ASA* ○ *NACB* ○ *NARAS* ○ *SMPTE* ○ *NAMBI*
○ *APRS* ○ *NAB* ○ *NAMM* ○ *SBE* ○ *SPARS* ○ *MEIEA*

Fulltime Faculty *2* **Parttime Faculty** *0*

Faculty Awards

Program Awards

Student Awards

Research areas

Indiana University

The Indiana University Audio Department offers two degrees in Audio Recording Technology, a two-year Associate of Science or a four-year Bachelor of Science. Both degrees offer training in audio recording and reinforcement with an emphasis toward classical music recording and producing. In addition, the AS degree emphasizes multitrack recording and media production, while the BS only offers one semester of multitrack recording. Applicants must demonstrate an interest in the activities of the School of Music. Performance ability is not required, though it is encouraged.

Classes are given in recording techniques, electronics, acoustics, maintenance, and musical styles. The School's facilities include digital recording, hard-disk editing, and analog multitrack recording.

The strengths of the Indiana University Audio Department are based on its interaction with the musical talent in the School of Music. Apart from formal classes in electronics and audio theory and practice, students have ample opportunity to experiment with recording a variety of high-caliber musical performances. These range on a weekly basis from fully-staged opera, symphony, and jazz concerts to daily student and faculty recitals. Recordings take place in the 1,460 seat Musical Arts Center and Recital Hall, both of which have fully-equipped recording facilities. Student recordings have appeared on National Public Radio, BBC Radio 3, and the Music from Indiana series. New recording facilities, involving a small, 100-seat hall and a medium, 750-seat hall are being constructed.

The Indiana University Audio Recording Technology programs stress musical values and positive interaction with musicians. Practical experience is given in recording sessions involving all types of music. The School regularly publishes CDs from the Audio Department's recordings.

In addition to the Audio Recording Technology programs, Indiana University also offers undergraduate and graduate degrees in composition with an emphasis in electronic and computer music. These programs utilize the School's Center for Electronic and Computer Music. Established in 1966, the Center houses two studios which employ the latest technologies in digital sound synthesis and sampling, MIDI, digital recording and editing, video, and research-level computing. The curriculum provides extensive training and historical background for students with little or no previous experience. More advanced students can enroll to use the studio facilities for the production of compositions and multi-media works, as well as for research.

Institute of Audio Research

Address	*64 University Place*		
	New York	*NY*	
	10003	*USA*	
Phone	*212-677-7580*	**FAX**	*212-677-6549*
Director	*Miriam Friedman*	**Admission Contact**	*Mark Khan*
Program Founded	*1969*	**School Type**	*Trade School*
Program Offered	*Recording Engineering and Production*		
Degrees Offered	*Diploma*		
Program Length	*7 months*		
Estimated Tuition	*$6,688*		
Main Emphasis	*Audio Engineering*	**Program is**	*Technical*
Accreditations	*ACTTS, NYS Ed Dept*		
Number of Studios	*2*	**Is school non-profit?**	*No*

Types of Studios
○ *Acoustic Research* ○ *Electronic Music* ○ *Film/Foley* ○ *Radio*
○ *Audio Research* ● *Music Recording* ○ *Television* ○ *Video*

Types of Recording
● *Analog Multitrack* ● *MIDI Sequencing* ○ *Video*
● *Digital Multitrack* ○ *DAW* ○ *Film*

Other Resources
● *Professional Studios* ○ *Television Stations*
○ *Radio Stations* ○ *Theater Tech Dept*

Class Size Lecture *30* **Class Size Lab** *15*

Assistance
● *Housing* ○ *Scholarships* ● *Internships*
● *Financial Aid* ○ *Work-study* ● *Job Placement*

Admission Policy *Selective* **Language** *English*

Prerequisites
● *High School Diploma* ○ *SAT-ACT* ○ *Music Audition*

Industry Affiliations
● *AES* ○ *ASA* ○ *NACB* ○ *NARAS* ● *SMPTE* ○ *NAMBI*
○ *APRS* ○ *NAB* ○ *NAMM* ○ *SBE* ● *SPARS* ● *MEIEA*

Fulltime Faculty *6* **Parttime Faculty** *20*

Faculty Awards

Program Awards

Student Awards

Research areas

Institute of Audio Research

The Institute of Audio Research's 650-hour Recording Engineering and Production (REP) program offers both the in-depth technical knowledge and hands-on skills required for an entry-level position as an assistant recording engineer or audio technician. The program is designed to give students practical working experience recording professional bands in state-of-the-art facilities- from the initial tracking to the final mix. Along with a strong theoretical background, special emphasis is placed on cutting edge music technology.

Students have interned at famous New York recording studios like Power Station, The Hit Factory, and Platinum Island. With minor adjustments in focus students can work effectively in other audio settings as well, including audio for film, video, and television, post-production, live sound, broadcast audio, satellite communications, corporate communications, and other areas.

In addition to the REP program, IAR offers a 930-hour Video Technology Program (VTP), as well as a series of short courses aimed at meeting the needs of working professionals in the audio and video industries. The school also provides custom training programs for media-related companies and manufacturers. For example, IAR is the US training facility for Solid State Logic (SSL), offering a series of hands-on seminars that range from practical operational skills to in-depth maintenance for consoles and automation software.

Institute of Communication Arts

Address *3246 Beta Avenue*
Burnaby *British Columbia*
V5G 4K4 *Canada*

Phone *604-298-5400* **FAX** *604-298-5403*

Director *Neils Nielsen* **Admission Contact** *Shannon Barker*

Program Founded *1979* **School Type** *Trade School*

Program Offered *Recording Arts, Interactive Multimedia Production*

Degrees Offered *Certificate*

Program Length *1-3 years*

Estimated Tuition *$6,000 per year*

Main Emphasis *Audio Engineering* **Program is** *Semi-technical*

Accreditations *BC Ministry of Advanced Education*

Number of Studios *6* **Is school non-profit?** *No*

Types of Studios ○ *Acoustic Research* ◉ *Electronic Music* ◉ *Film/Foley* ◉ *Radio*
○ *Audio Research* ◉ *Music Recording* ○ *Television* ◉ *Video*

Types of Recording ◉ *Analog Multitrack* ◉ *MIDI Sequencing* ◉ *Video*
◉ *Digital Multitrack* ◉ *DAW* ○ *Film*

Other Resources ◉ *Professional Studios* ○ *Television Stations*
◉ *Radio Stations* ○ *Theater Tech Dept*

Class Size Lecture *20* **Class Size Lab** *2*

Assistance ○ *Housing* ◉ *Scholarships* ◉ *Internships*
◉ *Financial Aid* ◉ *Work-study* ◉ *Job Placement*

Admission Policy *Selective* **Language** *English*

Prerequisites ◉ *High School Diploma* ○ *SAT-ACT* ○ *Music Audition*

Industry Affiliations ○ *AES* ○ *ASA* ○ *NACB* ○ *NARAS* ○ *SMPTE* ○ *NAMBI*
○ *APRS* ○ *NAB* ○ *NAMM* ○ *SBE* ○ *SPARS* ○ *MEIEA*

Fulltime Faculty *3* **Parttime Faculty** *15*

Faculty Awards

Program Awards

Student Awards

Research areas

Institute of Communication Arts

The Institute of Communication Arts (ICA) offers a variety of programs in media, including several in audio/music recording, music production and composition, interactive multi-media, and video. These programs vary in length from one to three years and are supported by scholarships and grants. The school's goal is to graduate media professionals.

Classes are offered in recording production, studio operations, audio electronics, sound reinforcement, audio for video, digital audio techniques, electronic music, video production, computer graphics, music theory and composition, and music business. These are supported by a variety of audio and video recording facilities.

ICA provides significant hands-on training with extensive experience in the school's studio production facilities. Training begins with comprehensive study and practical examples. Students then take control while instructors evaluate and refine the students' newly acquired skills. Students are allowed to work independently in a professional working environment while their progress is assessed. Graduates and senior students may apply to become Session Supervisors, allowing them to earn additional studio access in exchange for assisting newer students in facility operations. Students may also participate in routine maintenance, facility upgrades, and other special projects.

Courses at ICA may be taken on an individual/parttime basis, subject to availability. This is designed for employed students and for students wishing to take courses while deciding on a career path. Classes are offered evenings, days, and weekends. Semesters are fourteen weeks in length, and regular courses commence in February, June, and October.

International Film & Television Workshops

Address	*2 Central Street* *Rockport* *04856*	*ME* *USA*	
Phone	*207-236-8581*	**FAX**	*207-236-2558*
Director	*David Lyman*	**Admission Contact**	*Jennifer Seidel*
Program Founded	*1973*	**School Type**	*Workshop*
Program Offered	*Sound Recording, Video, Film, and Others*		
Degrees Offered	*None*		
Program Length	*3 days to 3 months*		
Estimated Tuition	*Varies*		
Main Emphasis	*Film/Video Production*	**Program is**	*Semi-technical*
Accreditations			
Number of Studios	*5*	**Is school non-profit?**	*No*
Types of Studios	○ *Acoustic Research* ◉ *Electronic Music* ◉ *Film/Foley* ○ *Radio* ○ *Audio Research* ◉ *Music Recording* ◉ *Television* ◉ *Video*		
Types of Recording	◉ *Analog Multitrack* ◉ *MIDI Sequencing* ◉ *Video* ◉ *Digital Multitrack* ◉ *DAW* ◉ *Film*		
Other Resources	◉ *Professional Studios* ○ *Television Stations* ○ *Radio Stations* ◉ *Theater Tech Dept*		
Class Size Lecture	*12-26*	**Class Size Lab**	*12-26*
Assistance	◉ *Housing* ◉ *Scholarships* ◉ *Internships* ◉ *Financial Aid* ◉ *Work-study* ○ *Job Placement*		
Admission Policy	*Highly selective*	**Language**	*English*
Prerequisites	○ *High School Diploma* ○ *SAT-ACT* ○ *Music Audition* *Professional Experience*		
Industry Affiliations	○ *AES* ○ *ASA* ○ *NACB* ○ *NARAS* ○ *SMPTE* ○ *NAMBI* ○ *APRS* ○ *NAB* ○ *NAMM* ○ *SBE* ○ *SPARS* ○ *MEIEA*		
Fulltime Faculty	*Varies*	**Parttime Faculty**	*Varies*
Faculty Awards			
Program Awards			
Student Awards			
Research areas			

International Film & Television Workshops

The International Film and Television Workshops offer a broad spectrum of over 100 courses related to sound, video, film, and other media. Programs cover a variety specialized studies aimed at enhancing the knowledge and experience of working professionals. Some of these courses are designed for directors, writers, producers, actors, editors, and camera operators while others deal with sound and music, professional video production, film and television documentary, television news and commercial production, corporate video, and personal and professional development. Courses vary in length from a few days to a couple weeks to three months. The workshops run from the end of May through the beginning of November. The programs are intended for students with professional experience as instruction is at advanced levels.

The workshops are held in Rockport, Maine, a remote village on the western edge of Penobscot Bay on the rugged coast of the North Atlantic. Here photographers, filmmakers, actors, writers, artists, and craftsmen of all kinds are able to exchange ideas, free from the turmoil and hectic pace of the average city. The Workshops' campus is located in several buildings scattered throughout Rockport. Facilities include classrooms, a 150-seat theater, a library, a book and supply store, a film soundstage and production studios, photography labs, and post-production center. Faculty, staff, and participants also benefit from sharing meals together at The Homestead, a 19th century farmhouse set on 20 acres just outside of Rockport.

Four workshops are offered that deal specifically with sound: The Location Sound Recording Workshop, Designing the Soundtrack, Scoring for Film and Videos: The Electronic Studio, and The Music Video Workshop. The Location Sound Recording Workshop is a one-week program in the techniques and tools used for location sound recording for feature films, documentaries, and videos. Designing the Soundtrack is a one-week course in how to create, design, budget and produce effective soundtracks for feature films, documentaries, and corporate video. Scoring for Film and Videos is a weekend workshop in learning how to construct a music soundtrack utilizing a professional recording studio in New York City. The Music Video Workshop is a one-week course in producing, directing, and making music videos.

Specialized workshops are also available for film and video technology covering such courses as EFP Steadicam, High Definition Television Camera, Underwater Video Camera, and 2-D and 3-D Animation. New courses are added regularly that reflect changes in media technology.

The faculty are all working professionals who take time from their professional lives to share their expertise, techniques, and experience with participants. The Workshop also offers employment opportunities for those interested in acting as instructors, teaching assistants, unit managers, studio and darkroom assistants, media technicians, drivers, and administrative assistants. Part-time and full-time positions are available, as well as paid and non-paid situations.

James Madison University

Address	*Music Dept* *Harrisonburg* *22807*	*VA* *USA*	
Phone	*703-568-6197*	**FAX**	*703-568-7819*
Director	*Dona Gilliam*	**Admission Contact**	*Mellasenah Morris*
Program Founded	*1977*	**School Type**	*University*
Program Offered	*Music Industry*		
Degrees Offered	*Bachelor of Music*		
Program Length	*4 years*		
Estimated Tuition	*$1788-$3620 per year*		
Main Emphasis	*Music Business*	**Program is**	*Semi-technical*
Accreditations	*NASM*		
Number of Studios	*3*	**Is school non-profit?**	*Yes*

Types of Studios
- ○ *Acoustic Research* ◉ *Electronic Music* ○ *Film/Foley* ○ *Radio*
- ○ *Audio Research* ◉ *Music Recording* ◉ *Television* ◉ *Video*

Types of Recording
- ◉ *Analog Multitrack* ◉ *MIDI Sequencing* ◉ *Video*
- ◉ *Digital Multitrack* ◉ *DAW* ○ *Film*

Other Resources
- ◉ *Professional Studios* ◉ *Television Stations*
- ◉ *Radio Stations* ○ *Theater Tech Dept*

Class Size Lecture *25* **Class Size Lab** *12*

Assistance
- ◉ *Housing* ◉ *Scholarships* ◉ *Internships*
- ◉ *Financial Aid* ◉ *Work-study* ◉ *Job Placement*

Admission Policy *Highly selective* **Language** *English*

Prerequisites ◉ *High School Diploma* ◉ *SAT-ACT* ◉ *Music Audition*

Industry Affiliations
- ◉ *AES* ○ *ASA* ○ *NACB* ◉ *NARAS* ○ *SMPTE* ◉ *NAMBI*
- ○ *APRS* ◉ *NAB* ◉ *NAMM* ○ *SBE* ○ *SPARS* ◉ *MEIEA*

Fulltime Faculty *35* **Parttime Faculty** *10*

Faculty Awards

Program Awards

Student Awards

Research areas

James Madison University

The JMU Department of Music is part of the College of Fine Arts and Communication and is housed in a state-of-the-art facility that was complete in 1989. Students in the Music Industry program have access to: a digital recording studio, a MIDI laboratory, an electronic music studio, and a music library, listening, and computer lab that houses MIDI workstations. Four rehearsal halls in the music building are connected to the control of the main recording studio to facilitate satellite recording of large ensembles. Studios are equipped with a variety of synthesizers, sound processing equipment, computers, and software to accommodate analog and digital recording needs. Students studying and researching legal aspects of the music industry also have access to the Law Library, which is housed in the Carrier Library on the JMU campus.

The JMU Music Industry program is committed to academic excellence and supports the University goal of integrating a broad liberal arts component into the student's major. The liberal studies component of the degree represents almost one-third of the course load required for graduation. It is designed to foster intellectual self-reliance as well as flexibility to think beyond the limits of professional and vocational study. Students wishing to become music industry majors must be accepted for admission to the University and pass audition requirements for majors in the Department of Music. The rigorous academic component of the program and its focus on music industry courses as part of a broader university sets JMU apart from other vocational/technical programs of study.

Keele University

Address	*Electronic Music & Recording Studios* *Keele* *Staffs.* *ST5 5BG* *UK*		
Phone	*0782-621111*	**FAX**	*0782-613847*
Director	*M. Vaughan/R.*	**Admission Contact**	*E. F. Slade*
Program Founded	*1989*	**School Type**	*University*
Program Offered	*Electronic Music, Digital Music Technology*		
Degrees Offered	*BA/BSc, MSc/MA/Diploma*		
Program Length	*3-4 years, 1 year*		
Estimated Tuition	*£6,000 , £2,000*		
Main Emphasis	*Electronic Music*	**Program is**	*Semi-technical*
Accreditations			
Number of Studios	*3*	**Is school non-profit?**	*No*
Types of Studios	○ *Acoustic Research* ◉ *Electronic Music* ○ *Film/Foley* ○ *Radio* ○ *Audio Research* ◉ *Music Recording* ○ *Television* ○ *Video*		
Types of Recording	◉ *Analog Multitrack* ◉ *MIDI Sequencing* ○ *Video* ○ *Digital Multitrack* ◉ *DAW* ○ *Film*		
Other Resources	○ *Professional Studios* ○ *Television Stations* ○ *Radio Stations* ○ *Theater Tech Dept*		
Class Size Lecture	*15*	**Class Size Lab**	*8*
Assistance	○ *Housing* ○ *Scholarships* ○ *Internships* ○ *Financial Aid* ○ *Work-study* ○ *Job Placement*		
Admission Policy		**Language**	*English*
Prerequisites	○ *High School Diploma* ○ *SAT-ACT* ○ *Music Audition* *A-level Music, BA/BSc for DMT*		
Industry Affiliations	○ *AES* ○ *ASA* ○ *NACB* ○ *NARAS* ○ *SMPTE* ○ *NAMBI* ◉ *APRS* ○ *NAB* ○ *NAMM* ○ *SBE* ○ *SPARS* ○ *MEIEA*		
Fulltime Faculty	*5*	**Parttime Faculty**	*1*
Faculty Awards			
Program Awards			
Student Awards			
Research areas	*Composition, Digital Music Technology, Aesthetics, English and American*		

Keele University

Keele University is a British University located 45 minutes from Manchester and Birmingham and 2 hours from London by rail. The University offers two programs of interest: an undergraduate degree in Electronic Music with an additional concentration and a graduate degree in Digital Music Technology. The University offers Joint Honors degrees in 34 subjects and hosts students from over 60 countries.

The Electronic Music program at Keele provides students with an opportunity to balance traditional areas of musical study with intensive training in music technology. The modular base of the program allows students some flexibility in structuring their studies so they can specialize in areas such as composition, recording techniques, or performance.

Most Keele students study two principal subjects. In combination with Electronic Music, students may specialize in American Studies, Biochemistry, Classical Studies, Computer Science, Criminology, Economics, Electronics, English, Environmental Management, Geography, Geology, German, History, Human Resources Management, International History, International Politics, Latin, Law, Mathematics, Physics, Politics, Russian Studies, Sociology, or a Double Language Combination. Many of these combinations are unique to Keele, and while the obvious combinations are with Electronics, Physics, Mathematics, or Computer Science, combinations with a language or social science are uniquely attractive. Career possibilities for graduates include technical work in broadcasting, recording, or software design, as well as original composition or studio work.

The graduate program in Digital Music Technology results in either a MSc, MA, or Diploma. The one-year course's aim is advanced multidisciplinary training for those interested in operating at a highly-specialized professional level in the music and studio industry as innovative product initiators/managers, senior electronic media executives, or audio software developers. The program covers a variety of topics, including electronics, signal theory, music theory and analysis, C programming, electro-acoustic composition, digital synthesis and signal processing, and advanced MIDI programming. A graduate project in either composition or musical software/development is also required.

Lebanon Valley College of Pennsylvania

Address *101 North College Avenue*
Annville *PA*
17003-0501 *USA*

Phone *717-867-6181* **FAX** *717-867-6124*

Director *Mark Mecham* **Admission Contact** *Mark Mecham*

Program Founded *1982* **School Type** *University*

Program Offered *Sound Recording Technology*

Degrees Offered *Bachelor of Music*

Program Length *4 years*

Estimated Tuition *$52,000*

Main Emphasis *Music Recording* **Program is** *Technical*

Accreditations *NASM*

Number of Studios *2* **Is school non-profit?** *Yes*

Types of Studios
○ *Acoustic Research* ○ *Electronic Music* ○ *Film/Foley* ○ *Radio*
○ *Audio Research* ◉ *Music Recording* ○ *Television* ○ *Video*

Types of Recording
◉ *Analog Multitrack* ○ *MIDI Sequencing* ○ *Video*
○ *Digital Multitrack* ○ *DAW* ○ *Film*

Other Resources
○ *Professional Studios* ○ *Television Stations*
○ *Radio Stations* ○ *Theater Tech Dept*

Class Size Lecture *12* **Class Size Lab** *5*

Assistance
○ *Housing* ◉ *Scholarships* ◉ *Internships*
◉ *Financial Aid* ◉ *Work-study* ◉ *Job Placement*

Admission Policy *Selective* **Language** *English*

Prerequisites ◉ *High School Diploma* ◉ *SAT-ACT* ◉ *Music Audition*

Industry Affiliations
◉ *AES* ○ *ASA* ○ *NACB* ○ *NARAS* ○ *SMPTE* ○ *NAMBI*
○ *APRS* ○ *NAB* ○ *NAMM* ○ *SBE* ◉ *SPARS* ○ *MEIEA*

Fulltime Faculty *1* **Parttime Faculty** *2*

Faculty Awards

Program Awards

Student Awards

Research areas

Lebanon Valley College of Pennsylvania

Lebanon Valley College offers a Bachelor of Music with an emphasis in Sound Recording Technology. The program benefits from hands-on experience in two multitrack control rooms, providing students with ample individual studio experience. The music program features a student-centered curriculum and individualized attention. The Sound Recording Technology emphasis is balanced between music, sound recording, science, math, physics, and liberal arts courses. Students are also required to record and/or perform within a variety of large and small ensembles.

Knowing about music, developing sensitive and critical listening skills, creating and performing music, and understanding the role of music in history and contemporary society are all vital elements in the music curriculum at Lebanon Valley. A broad-based, flexible course of study is designed to meet specific professional requirements of the music student as well as the musical needs of the general liberal arts student. Many graduates immediately embark on music industry careers.

Los Angeles City College

Address	*855 North Vermont Avenue* *Los Angeles* *CA* *91326* *USA*		
Phone	*213-953-4521*	**FAX**	
Director	*J. Robert Stahley*	**Admission Contact**	*Myra Siegel*
Program Founded	*1930*	**School Type**	*Community College*
Program Offered	*Radio Broadcasting/Recording Arts,Television,Cinema*		
Degrees Offered	*Associate of Arts*		
Program Length	*2 years*		
Estimated Tuition	*$600*		
Main Emphasis	*Broadcasting*	**Program is**	*Semi-technical*
Accreditations	*Western States*		
Number of Studios	*15*	**Is school non-profit?**	*Yes*
Types of Studios	○ *Acoustic Research* ○ *Electronic Music* ○ *Film/Foley* ● *Radio* ○ *Audio Research* ● *Music Recording* ○ *Television* ● *Video*		
Types of Recording	● *Analog Multitrack* ○ *MIDI Sequencing* ● *Video* ○ *Digital Multitrack* ○ *DAW* ● *Film*		
Other Resources	● *Professional Studios* ○ *Television Stations* ● *Radio Stations* ● *Theater Tech Dept*		
Class Size Lecture	*35*	**Class Size Lab**	*30*
Assistance	○ *Housing* ● *Scholarships* ○ *Internships* ● *Financial Aid* ● *Work-study* ○ *Job Placement*		
Admission Policy	*Open*	**Language**	*English*
Prerequisites	● *High School Diploma* ○ *SAT-ACT* ○ *Music Audition*		
Industry Affiliations	○ *AES* ○ *ASA* ○ *NACB* ○ *NARAS* ● *SMPTE* ○ *NAMBI* ○ *APRS* ● *NAB* ○ *NAMM* ○ *SBE* ○ *SPARS* ○ *MEIEA*		
Fulltime Faculty	*9*	**Parttime Faculty**	*6*
Faculty Awards			
Program Awards			
Student Awards	*AFI Grants, MTV Awards, Emmy/Grammy Nominees, ACE Awards,*		
Research areas			

Los Angeles City College

The Radio-Television-Film Department at Los Angeles City College offers students comprehensive media training. The programs provide basic and advanced courses in the latest studio and field production techniques using state-of-the-art equipment and facilities. With no prior experience required, students begin hands-on production in their first semester. They then advance to courses that prepare them for employment in the industry. The LACC Communication Center houses all three programs in one centralized location.

For Radio Broadcasting/Recording Arts, there are post-production facilities, including 18 radio production rooms, for producing shows for the schools on-air broadcast station. The Recording Arts program utilizes multitrack studios for recording live bands.

For Cinema, there is a fully-equipped film sound stage, 26 editing rooms, 3 screening rooms, plus full audio post-production and animation facilities. In addition, there is extensive 16mm production equipment available for all students.

For Television, there are two broadcast-quality television studios. A comprehensive master-control interconnects the studios as well as an off-line video editing suite. Remote equipment is available for electronic field production and news gathering.

The school's faculty and staff consists of over two-dozen professional, engineers, and educators who provide expert technical advice combined with detailed information on current industry procedures. Graduates from LACC are active at all levels in the industry and have won a variety of awards.

Loyola Marymount University

Address *Communication Arts Dept*
Los Angeles *CA*
90045 *USA*

Phone *310-338-3033* **FAX** *310-338-3030*

Director *John Michael* **Admission Contact** *Patricia Oliver*

Program Founded *1985* **School Type** *University*

Program Offered *Recording Arts*

Degrees Offered *Bachelor of Arts*

Program Length *4 years*

Estimated Tuition *$48,000*

Main Emphasis *Audio Engineering* **Program is** *Technical*

Accreditations

Number of Studios *6* **Is school non-profit?** *Yes*

Types of Studios
- ○ *Acoustic Research* ○ *Electronic Music* ◉ *Film/Foley* ○ *Radio*
- ○ *Audio Research* ◉ *Music Recording* ◉ *Television* ○ *Video*

Types of Recording
- ◉ *Analog Multitrack* ○ *MIDI Sequencing* ◉ *Video*
- ○ *Digital Multitrack* ○ *DAW* ◉ *Film*

Other Resources
- ○ *Professional Studios* ○ *Television Stations*
- ○ *Radio Stations* ○ *Theater Tech Dept*

Class Size Lecture *14* **Class Size Lab** *14*

Assistance
- ○ *Housing* ◉ *Scholarships* ◉ *Internships*
- ◉ *Financial Aid* ◉ *Work-study* ○ *Job Placement*

Admission Policy *Highly selective* **Language** *English*

Prerequisites ◉ *High School Diploma* ◉ *SAT-ACT* ○ *Music Audition*

Industry Affiliations
- ◉ *AES* ○ *ASA* ○ *NACB* ○ *NARAS* ○ *SMPTE* ○ *NAMBI*
- ○ *APRS* ○ *NAB* ○ *NAMM* ○ *SBE* ○ *SPARS* ○ *MEIEA*

Fulltime Faculty *11* **Parttime Faculty** *Varies*

Faculty Awards

Program Awards

Student Awards

Research areas

Loyola Marymount University

The LMU Recording Arts program offers students the opportunity to explore both the aesthetic and technical challenges of sound design and recording through in-depth study and hands-on experience. Emphasis is placed on both the art of music recording and the creative use of sound in film and television. Students take classes in mass communications, cinema theory and history, screenwriting, film and television production, sound design, recording technology, acoustics, production and post-production sound, studio recording practices and techniques, digital audio theory, advanced recording, and related music courses. Among LMU's recording facilities are a film re-recording studio, a video-assisted film and television post-production suite, and a fully-equipped multitrack recording studio. Classes are kept small to insure that the needs of individual students are met.

The Communication Arts Department at LMU also offers programs in writing for film and television, film and television production, and communications studies. All students benefit from a variety of production facilities, including: a cinema sound stage, a television studio, several post-production studios, four sound recording studios, four screening rooms, an animation studio, and a FM stereo radio station. The radio station, KXLU, was recently cited as one of the leading student radio stations in the country. LMU students have won national and international film and writing awards and have gone on to key positions in the entertainment and communications industries.

McGill University, Faculty of Music

Address *555 Sherbrooke Street West*
Montreal *Quebec*
H3A 1E3 *Canada*

Phone *514-398-4535* **FAX** *514-398-8061*

Director *Wieslaw Woszczyk* **Admission Contact** *John Grew*

Program Founded *1979* **School Type** *University*

Program Offered *Sound Recording*

Degrees Offered *Master of Music*

Program Length *2-3 years*

Estimated Tuition *$1,350-$7,600 per year*

Main Emphasis *Music Recording* **Program is** *Technical*

Accreditations

Number of Studios *5* **Is school non-profit?** *Yes*

Types of Studios ○ *Acoustic Research* ○ *Electronic Music* ○ *Film/Foley* ○ *Radio* ○ *Audio Research* ○ *Music Recording* ○ *Television* ○ *Video*

Types of Recording ◉ *Analog Multitrack* ○ *MIDI Sequencing* ○ *Video* ◉ *Digital Multitrack* ◉ *DAW* ○ *Film*

Other Resources ◉ *Professional Studios* ○ *Television Stations* ○ *Radio Stations* ○ *Theater Tech Dept*

Class Size Lecture *4-5* **Class Size Lab** *4-5*

Assistance ○ *Housing* ○ *Scholarships* ○ *Internships* ○ *Financial Aid* ○ *Work-study* ○ *Job Placement*

Admission Policy *Highly selective* **Language** *English*

Prerequisites ○ *High School Diploma* ○ *SAT-ACT* ○ *Music Audition*
Bachelor of Music

Industry Affiliations ◉ *AES* ○ *ASA* ○ *NACB* ○ *NARAS* ○ *SMPTE* ○ *NAMBI* ○ *APRS* ○ *NAB* ○ *NAMM* ○ *SBE* ○ *SPARS* ○ *MEIEA*

Fulltime Faculty **Parttime Faculty**

Faculty Awards

Program Awards

Student Awards

Research areas

McGill University, Faculty of Music

McGill University's Music Department offers a Master of Music in Sound Recording degree. The program can be taken with or without a thesis. The non-thesis option includes a variety of courses, including: Technical Ear Training, Digital Studio Technology, Digital/Analog Audio Editing, Special Topics in Classical Music Recording, Music Recording Theory and Practice, Analysis of recording, Studio Equipment Maintenance Theory and Practice, Sound with Vision, Advanced Digital Editing and Post-Production, and others. The thesis option includes: Technical Ear Training, Music Theory and Practice, Analysis of Recordings, Studio Equipment Maintenance, and a written thesis and recording presentation. Thesis candidates undertake supervised research leading to a thesis involving an in-depth investigation of a specialized field of music recording theory or practice, along with a presentation of one hour of their recording projects.

Students are able to take advantage of three recording halls, four control rooms, and one two-room studio. Facilities include multitrack digital recording and editing, computer-assisted mixing, and a electronic repair shop.

The program is highly selective and only admits four to five students per year. Non-McGill graduates are required to take a year of prerequisite courses in mathematics, psychophysics of music, music acoustics, orchestration, electro-acoustic music, electronics, music recording, computer applications in music, and electro-acoustics.

McLennan Community College

Address	*1400 College Drive*		
	Waco	*TX*	
	76708	*USA*	
Phone	*817-750-3578*	**FAX**	*817-756-0934*
Director	*David Hibbard*	**Admission Contact**	*David Hibbard*
Program Founded	*1980*	**School Type**	*Community College*
Program Offered	*Commercial Music with an emphasis in Audio Technology*		
Degrees Offered	*Associate of Applied Sciences*		
Program Length	*2 years*		
Estimated Tuition	*$500-$1500*		
Main Emphasis	*Music Recording*	**Program is**	*Semi-technical*
Accreditations			
Number of Studios	*2*	**Is school non-profit?**	*Yes*

Types of Studios
- O *Acoustic Research* O *Electronic Music* O *Film/Foley* O *Radio*
- O *Audio Research* ◉ *Music Recording* O *Television* O *Video*

Types of Recording
- ◉ *Analog Multitrack* ◉ *MIDI Sequencing* O *Video*
- O *Digital Multitrack* O *DAW* O *Film*

Other Resources
- O *Professional Studios* O *Television Stations*
- O *Radio Stations* O *Theater Tech Dept*

Class Size Lecture *14* **Class Size Lab** *12*

Assistance
- O *Housing* ◉ *Scholarships* O *Internships*
- ◉ *Financial Aid* ◉ *Work-study* O *Job Placement*

Admission Policy *Open* **Language** *English*

Prerequisites
- ◉ *High School Diploma* ◉ *SAT-ACT* ◉ *Music Audition*
- *Interview*

Industry Affiliations
- O *AES* O *ASA* O *NACB* O *NARAS* O *SMPTE* O *NAMBI*
- O *APRS* O *NAB* O *NAMM* O *SBE* O *SPARS* ◉ *MEIEA*

Fulltime Faculty *10* **Parttime Faculty** *2*

Faculty Awards

Program Awards

Student Awards

Research areas

McLennan Community College

McLennan Community College's AAS in Audio Technology degree is one of three programs offered in their Commercial Music area. Current enrollment in the program are 100 students from throughout the United States. The Commercial Music program has 20 ensembles performing the musical styles of rock, jazz, country, soul, and contemporary Christian. The college also offers students educational opportunities in traditional music areas including band, choir, opera, and chamber ensembles.

The program is housed in the school's Performing Arts Center and benefits from a multitrack studio and fully-equipped MIDI facility. The audio technology curriculum requires the development of skills in performance, composition, and management in addition to the areas of production, engineering, and maintenance. As part of the program, students must successfully plan, develop, record, and manage their own recording projects.

McLennan Community College is situated on a 190 acre campus within the city limits of Waco, Texas and is adjacent to Cameron Park and Lake Brazos. Total enrollment is over 6,000 full-time students, and another 12,000 students enroll in non-credit classes. The college offers equal educational opportunities and financial aid to all qualified students and does not discriminate on any basis.

Media Arts Center

Address	*753 Capitol Avenue*		
	Hartford	*CT*	
	06106	*USA*	
Phone	*203-951-8175*	**FAX**	
Director	*Jack Stang*	**Admission Contact**	
Program Founded	*1975*	**School Type**	*Trade School*
Program Offered	*Recording Engineering*		
Degrees Offered	*Certificate*		
Program Length	*6 weeks, 2 years*		
Estimated Tuition	*$1,700-$4,000*		
Main Emphasis	*Audio Engineering*	**Program is**	*Semi-technical*
Accreditations			
Number of Studios	*2*	**Is school non-profit?**	*No*

Types of Studios
O *Acoustic Research* O *Electronic Music* O *Film/Foley* O *Radio*
O *Audio Research* ◉ *Music Recording* O *Television* O *Video*

Types of Recording
◉ *Analog Multitrack* ◉ *MIDI Sequencing* O *Video*
O *Digital Multitrack* O *DAW* O *Film*

Other Resources
◉ *Professional Studios* O *Television Stations*
◉ *Radio Stations* O *Theater Tech Dept*

Class Size Lecture	*10-15*	**Class Size Lab**	*10-15*

Assistance
O *Housing* O *Scholarships* O *Internships*
O *Financial Aid* O *Work-study* O *Job Placement*

Admission Policy	*Selective*	**Language**	*English*

Prerequisites
◉ *High School Diploma* O *SAT-ACT* O *Music Audition*

Industry Affiliations
O *AES* O *ASA* O *NACB* O *NARAS* O *SMPTE* O *NAMBI*
O *APRS* O *NAB* O *NAMM* O *SBE* O *SPARS* O *MEIEA*

Fulltime Faculty	*4*	**Parttime Faculty**	*0*

Faculty Awards

Program Awards

Student Awards

Research areas

Media Arts Center

The Media Arts Center offers two part-time recording programs as well as individual evening classes. Their six-week Recording Engineering program is offered each summer with classes running Monday through Thursday, 9 am- 4 pm. The same program is offered over two years with classes meeting on Wednesday and Friday, 9:30 am- 4 pm. All classes are held in a full-time professional multitrack studio and are taught by professional working engineers.

The program encourages interaction between students and instructors, blending lecture and practical class time. The school aims to keep students active participants in the recording process. The curriculum consists of: Multitrack I & II, MIDI I & II, Problem Solving I & II, and Record Production I & II. These courses cover a wide range of topics and provide students with plenty of hands-on experience in the Media Arts Center's facilities.

Media Production Facilities

Address *Bon Marche Bldg., Ferndale Road*
London
SW9 8EJ *UK*

Phone *44-71-274-4000* **FAX** *44-71-738-5428*

Director *Simaen Skolfield* **Admission Contact** *Paul Halpin*

Program Founded **School Type** *Trade School*

Program Offered *Advanced Sound Recording & Production Techniques*

Degrees Offered *Diploma, Certificate*

Program Length *3 months- 1 year*

Estimated Tuition *£2,500- £6,000*

Main Emphasis *Audio Engineering* **Program is** *Technical*

Accreditations

Number of Studios *2* **Is school non-profit?** *No*

Types of Studios
- ○ *Acoustic Research* ◉ *Electronic Music* ○ *Film/Foley* ○ *Radio*
- ○ *Audio Research* ◉ *Music Recording* ○ *Television* ○ *Video*

Types of Recording
- ◉ *Analog Multitrack* ◉ *MIDI Sequencing* ○ *Video*
- ◉ *Digital Multitrack* ◉ *DAW* ○ *Film*

Other Resources
- ◉ *Professional Studios* ○ *Television Stations*
- ○ *Radio Stations* ○ *Theater Tech Dept*

Class Size Lecture *Varies* **Class Size Lab** *Varies*

Assistance
- ◉ *Housing* ○ *Scholarships* ○ *Internships*
- ○ *Financial Aid* ○ *Work-study* ○ *Job Placement*

Admission Policy *Selective* **Language** *English*

Prerequisites ○ *High School Diploma* ○ *SAT-ACT* ○ *Music Audition*

Industry Affiliations
- ◉ *AES* ○ *ASA* ○ *NACB* ○ *NARAS* ○ *SMPTE* ○ *NAMBI*
- ○ *APRS* ○ *NAB* ○ *NAMM* ○ *SBE* ○ *SPARS* ○ *MEIEA*

Fulltime Faculty *6* **Parttime Faculty** *6-10*

Faculty Awards *Grammys*

Program Awards

Student Awards

Research areas

Media Production Facilities

Media Production Facilities in London offers a program in Advanced Sound Recording and Production Techniques. The one-year Diploma course is a full-time program consisting of three 12-week modules: Analog Sound Recording and Production Techniques, Digital Sound Recording and Production Techniques, and Creative Music Recording and Production Techniques. The modules can be taken individually and commence January, May, and September of each year.

During the first module, Analog Sound Recording and Production Techniques, students develop practical skills in multitrack sound recording and mixing while acquiring a solid foundation in audio theory. Hands-on experience is gained through operating the program's professionally-equipped recording studio and digital programming suite. Students, working in small groups under the guidance of an experienced engineer, actively participate in all aspects of multitrack recording and mixing.

The second module, Digital Sound Recording and Production Techniques, emphasizes digital technology and its application in sound production, recording, and post-production. Students attend seminars and lectures with specialists in all areas of digital audio, including computer-based MIDI music production, automated sound mixing, and all forms of digital audio recording. Cooperation with various professional studios and audio manufacturers allows Media to provide in-house training on a variety of digital audio systems.

The third module, Creative Music Recording and Production Techniques, stresses practical experience and encourages students to creatively apply their newly acquired knowledge and skills. Professionals share tricks of the trade, and students develop their own music projects as well as those with bands and solo artists. Sessions at other professional recording facilities provide practical mixing and mastering experience at the highest level. An introduction to audio post-production techniques is also included. The program concludes with an examination of employment opportunities in the industry.

Media also offers a one-week, full-time Introduction to Sound Recording and Mixing program that is limited to a maximum of eight students. The course provides a basic introduction to multitrack studio operations, including mixing console signal flow, tape machine operation, microphone placment, MIDI programming, and mixing techniques. The tuition for this program is £295 + VAT.

Media employs three full-time engineers in addition to six regular lecturers and a pool of additional guest lecturers from the recording industry. Director Simaen Skolfield is a Grammy-award winning engineer with a broad background in the audio industry. A list of other active faculty members and guest lecturers reads like a who's who in audio and includes John Watkinson, Richard Salter, David Pope, Tony Waldron, Tony Faulkner, Andy Munro, Rupert Neve, Steve Smith, Karl Walters, and Andy Day.

Memphis State University

Address	*CFA 232*		
	Memphis	*TN*	
	38152	*USA*	
Phone	*901-678-2559*	**FAX**	*901-678-5118*
Director	*Larry Lipman*	**Admission Contact**	
Program Founded	*1981*	**School Type**	*University*
Program Offered	*Commercial Music*		
Degrees Offered	*Bachelor of Music*		
Program Length	*4 years*		
Estimated Tuition	*$6,400-20,000*		
Main Emphasis	*Audio Engineering*	**Program is**	*Semi-technical*
Accreditations	*NASM*		
Number of Studios	*1*	**Is school non-profit?**	*Yes*

Types of Studios
- ○ *Acoustic Research* ◉ *Electronic Music* ◉ *Film/Foley* ◉ *Radio*
- ○ *Audio Research* ◉ *Music Recording* ◉ *Television* ◉ *Video*

Types of Recording
- ◉ *Analog Multitrack* ◉ *MIDI Sequencing* ◉ *Video*
- ○ *Digital Multitrack* ◉ *DAW* ○ *Film*

Other Resources
- ◉ *Professional Studios* ◉ *Television Stations*
- ◉ *Radio Stations* ◉ *Theater Tech Dept*

Class Size Lecture	*16*	**Class Size Lab**	*16*

Assistance
- ◉ *Housing* ◉ *Scholarships* ◉ *Internships*
- ◉ *Financial Aid* ◉ *Work-study* ○ *Job Placement*

Admission Policy	*Selective*	**Language**	*English*

Prerequisites
- ◉ *High School Diploma* ◉ *SAT-ACT* ○ *Music Audition*

Industry Affiliations
- ◉ *AES* ○ *ASA* ○ *NACB* ◉ *NARAS* ○ *SMPTE* ○ *NAMBI*
- ○ *APRS* ○ *NAB* ○ *NAMM* ○ *SBE* ◉ *SPARS* ◉ *MEIEA*

Fulltime Faculty	*4*	**Parttime Faculty**	*4+*
Faculty Awards			
Program Awards			
Student Awards	*NARAS Collegiate Student Music Award*		
Research areas			

Memphis State University

Memphis State offers a Bachelor of Music in Commercial Music with concentrations in Recording Technology, Music Business, Jazz Composition, and Jazz Performance. The programs stress a through understanding of fundamental concepts, yet place equal emphasis on developing the student's ability to quickly adapt to new practices, technologies and creative directions. The program's instructors are actively involved in the commercial music industry and possess a broad knowledge of industry practices. CMUS majors have won the prestigious NARAS Collegiate Music Award.

The Commercial Music program benefits from on-campus production facilities, including a comprehensive multitrack recording studio, a synthesis suite, an extensive MIDI-based electronic music lab, and a complete video production facility. In addition to these facilities, students work in music, fine arts, and dance studios, as well as computer art, writing, and publishing labs. The facilities serve instructional needs as well as supporting the University's Highwater record label, its two music publishing companies, and video production activities. Commercial Music majors are encouraged to become active in Highwater productions.

The Memphis arts community offers a rich assortment of internship possibilities and diverse cultural opportunities.
Generous educational grants are provided by supporting organizations such as the Memphis chapter of NARAS and the Jazz Society of Memphis. MSU students have also won national scholarship competitions sponsored by the National Association of Jazz Educators, Downbeat magazine, and the Audio Engineering Society. Other scholarship funds are available for exceptional students, and many states offer financial assistance through the Academic Common Market program. The program's commitment to personal attention and quality instruction requires enrollment to be limited based on selective procedures.

Middle Tennessee State University

Address	*PO Box 21 MTSU*		
	Murfreesboro	*TN*	
	37132	*USA*	
Phone	*615-898-2578*	**FAX**	*615-898-5682*
Director	*Richard Barnet*	**Admission Contact**	*Lynn Palmer*
Program Founded	*1975*	**School Type**	*University*
Program Offered	*Recording Industry Management*		
Degrees Offered	*Bachelor of Science*		
Program Length	*4 years*		
Estimated Tuition	*$6,400*		
Main Emphasis	*Audio Engineering*	**Program is**	*Semi-technical*
Accreditations	*SACS*		
Number of Studios	*4*	**Is school non-profit?**	*Yes*

Types of Studios ○ *Acoustic Research* ◉ *Electronic Music* ○ *Film/Foley* ○ *Radio* ○ *Audio Research* ◉ *Music Recording* ○ *Television* ○ *Video*

Types of Recording ◉ *Analog Multitrack* ◉ *MIDI Sequencing* ○ *Video* ◉ *Digital Multitrack* ◉ *DAW* ○ *Film*

Other Resources ○ *Professional Studios* ◉ *Television Stations* ◉ *Radio Stations* ◉ *Theater Tech Dept*
Video Post-Production Dept.

Class Size Lecture *30* **Class Size Lab** *15*

Assistance ◉ *Housing* ◉ *Scholarships* ◉ *Internships* ◉ *Financial Aid* ◉ *Work-study* ◉ *Job Placement*

Admission Policy *Open* **Language** *English*

Prerequisites ◉ *High School Diploma* ◉ *SAT-ACT* ○ *Music Audition*
Application

Industry Affiliations ◉ *AES* ○ *ASA* ○ *NACB* ◉ *NARAS* ○ *SMPTE* ○ *NAMBI* ○ *APRS* ○ *NAB* ◉ *NAMM* ○ *SBE* ◉ *SPARS* ◉ *MEIEA*

Fulltime Faculty *13* **Parttime Faculty** *3*

Faculty Awards *2 Outstanding Teacher Awards*

Program Awards *TEC Nominee- 8 years*

Student Awards *NARAS Outstanding Student Music Award*

Research areas *Acoustics, Instructional Technology*

Middle Tennessee State University

The Recording Industry Management (RIM) degree program at MTSU offers students a variety of curriculums designed to prepare them for entry-level positions in all phases of the recording industry from record companies to music publishing firms, recording studios, talent agencies, concert promotions, record retail and distribution firms, manufacturing companies, audio consulting, and trade publications.

The program was designed with consultation and continuing advice from NARAS, NARM, SPARS, NMPA, and the AES. Degree requirements, in addition to the University's general studies requirements include a total of 36 credit hours of RIM courses, a minor in Mass Communications, and a second minor in either Business Administration, Electronics, Music Industry, or Entertainment Technology. There are an additional 16 credit hours of free electives, allowing students to further concentrate their studies in audio engineering, business, or music.

The RIM Department and School of Mass Communication are equipped to develop the technical skills necessary in the recording industry. A new $15 million dollar Mass Communications building was opened in January 1991. Students benefit from several on-campus recording studios used exclusively to train students in the techniques of analog and digital recording and mixing. Students can also take advantage of the school's 50,000 watt FM radio station for developing their broadcast production skills. In addition, there are professional-quality video studio, photography labs, AV equipment, and computer graphics equipment available to students.

MTSU houses the Center for Popular Music, an archive of popular recordings, sheet music, music video, and music-related publications. The Center is available for research and sponsorship of activities in the business and sociological aspects of popular music.

Located in Murfreesboro, MTSU is only 30 miles southeast of Nashville where students often go to learn more about the industry through internships, personal contacts, and the other activities of "Music City".

Midland College

Address *3600 N. Garfield*
Midland *TX*
79705 *USA*

Phone *915-685-4648* **FAX** *915-685-4714*

Director *Robert Hunt* **Admission Contact** *Robert Hunt*

Program Founded **School Type** *Community College*

Program Offered *Audio Technology*

Degrees Offered *Associate in Applied Science/Arts, Certificate*

Program Length *2 years, 1 year*

Estimated Tuition *$200 per semester*

Main Emphasis *Music Recording* **Program is** *Semi-technical*

Accreditations *Texas Association of Music Schools, NASM*

Number of Studios *1* **Is school non-profit?** *Yes*

Types of Studios
○ *Acoustic Research* ○ *Electronic Music* ○ *Film/Foley* ○ *Radio*
○ *Audio Research* ◉ *Music Recording* ○ *Television* ○ *Video*

Types of Recording
◉ *Analog Multitrack* ◉ *MIDI Sequencing* ○ *Video*
○ *Digital Multitrack* ○ *DAW* ○ *Film*

Other Resources
◉ *Professional Studios* ○ *Television Stations*
○ *Radio Stations* ○ *Theater Tech Dept*

Class Size Lecture *12* **Class Size Lab** *3*

Assistance
○ *Housing* ◉ *Scholarships* ○ *Internships*
◉ *Financial Aid* ◉ *Work-study* ○ *Job Placement*

Admission Policy *Open* **Language** *English*

Prerequisites ◉ *High School Diploma* ○ *SAT-ACT* ○ *Music Audition*

Industry Affiliations
○ *AES* ○ *ASA* ○ *NACB* ○ *NARAS* ○ *SMPTE* ○ *NAMBI*
○ *APRS* ○ *NAB* ○ *NAMM* ○ *SBE* ○ *SPARS* ○ *MEIEA*

Fulltime Faculty *1* **Parttime Faculty** *12*

Faculty Awards

Program Awards

Student Awards

Research areas

Midland College

The Midland College of Audio Technology program is designed to offer students both the theoretical background and hands-on experience necessary to prepare them for an entry-level position in professional audio production. The curriculum emphasizes music, electronics, and recording technology courses to provide a well-balanced program. Graduate receive an Associate in Applied Science degree with a major in Audio Technology. A special one-year certificate program is also available.

Recording classes are taught in Midland College's professional multitrack recording studio. The studio includes facilities for MIDI synthesis and sequencing as well as SMPTE time code equipment.

Midland College's Department of Music also offers an Associate in Arts degree with a specialization in Commercial Music. The program is designed to acquaint students with current trends in music production and performance. Students interested in this program consult with the faculty to develop a special individualized curriculums.

As a community service, the department of Music provides opportunities for adults and precollege students to participate in various courses, private lessons, and college-sponsored performing ensembles. Students planning to transfer to a particular university can arrange their programs to meet the requirements of the particular college.

Musikhögskolan i Piteå

Address *Box 210*
Piteå
941 25 *Sweden*

Phone *46-911-72627* **FAX** *46-911-72610*

Director *Lars Hallberg* **Admission Contact** *Lars Hallberg*

Program Founded *1991* **School Type** *University*

Program Offered *Sound Engineering*

Degrees Offered *University Certificate*

Program Length *2 years*

Estimated Tuition *Inquire*

Main Emphasis *Music Recording* **Program is** *Technical*

Accreditations

Number of Studios *2* **Is school non-profit?** *Yes*

Types of Studios ○ *Acoustic Research* ◉ *Electronic Music* ○ *Film/Foley* ◉ *Radio*
○ *Audio Research* ◉ *Music Recording* ○ *Television* ○ *Video*

Types of Recording ◉ *Analog Multitrack* ◉ *MIDI Sequencing* ○ *Video*
○ *Digital Multitrack* ○ *DAW* ○ *Film*

Other Resources ◉ *Professional Studios* ◉ *Television Stations*
◉ *Radio Stations* ◉ *Theater Tech Dept*
Swedish Broadcasting Company

Class Size Lecture *10* **Class Size Lab** *10*

Assistance ○ *Housing* ○ *Scholarships* ○ *Internships*
○ *Financial Aid* ○ *Work-study* ○ *Job Placement*

Admission Policy *Highly selective* **Language** *Swedish*

Prerequisites ◉ *High School Diploma* ○ *SAT-ACT* ○ *Music Audition*
Basic Electronics Background

Industry Affiliations ○ *AES* ○ *ASA* ○ *NACB* ○ *NARAS* ○ *SMPTE* ○ *NAMBI*
○ *APRS* ○ *NAB* ○ *NAMM* ○ *SBE* ○ *SPARS* ○ *MEIEA*

Fulltime Faculty *Varies* **Parttime Faculty** *Varies*

Faculty Awards

Program Awards

Student Awards

Research areas

Musikhögskolan i Piteå

The School of Music in Piteå, Sweden started the country's first Sound Engineering program in the fall of 1991. The two-year program is designed to train students interested in sound engineering, radio broadcasting, music recording, and theatrical sound design. The first year of studies covers the basic course in Sound Engineering and Radio Broadcasting. After completing this year, student can choose whether to enter the workforce as sound technicians or continue their studies to earn a University Certificate in Sound Engineering.

The second year of study offers two specializations: Theatrical Sound Engineering or Music Recording. If the school has vacancies in the second year, working sound technicians, who meet the prerequisites, can be accepted into the program. Acceptance into the program is very competitive as only ten students are admitted per year. Applicants should have two years of English, an educational background in electronics or electrotechnical program, or a comparable education, along with documented musical activity in both municipal music school and in ensemble performance. Good hearing, documented by a doctor, is also required. Qualified applicants that are selected will be given a practical test and an interview for formal admission into the program.

The Sound Engineering program is housed in a specially-designed building that adjoins the School of Music. It contains classrooms, two studios and control rooms, and a computer music facility. There are also lines that tie the Sound Engineering School to the School of Music's facilities, including their electro-acoustic music studio, main auditorium, and organ concert hall. The concert hall is acoustically similar to a church and contains a large, 34-stop organ. The school also benefits from a MIDI grand piano in the main studio.

The school's main studio is a large music recording studio built for acoustic music, but also well-suited for rock music and theatrical recording. The adjacent music control room is large enough to teach students in a variety of recording situations. The room also has a workbench with earphone listening stations for each student. The second control room and studio were built for speech, corresponding to a Swedish Broadcasting Company transmission block. Recordings from the main studio can also be made in the second control room, given the flexible layout of the facility.

The deadline for applications is April 1st of each year.

New York University, Department of Music

Address	*34 West 4th Street*		
	New York	*NY*	
	10003	*USA*	
Phone	*212-998-5422*	**FAX**	*212-995-4043*
Director	*Kenneth Peacock*	**Admission Contact**	*Kenneth Peacock*
Program Founded	*1988*	**School Type**	*University*
Program Offered	*Music Technology, Music Business*		
Degrees Offered	*Bachelor of Music, Master of Music*		
Program Length	*BM- 4 years, MM- 2 years*		
Estimated Tuition	*$15,000-$25,000*		
Main Emphasis	*Electronic Music*	**Program is**	*Semi-technical*
Accreditations	*NASM*		
Number of Studios	*10*	**Is school non-profit?**	*Yes*

Types of Studios
- ◉ *Acoustic Research* ◉ *Electronic Music* ○ *Film/Foley* ○ *Radio*
- ◉ *Audio Research* ◉ *Music Recording* ○ *Television* ◉ *Video*

Types of Recording
- ◉ *Analog Multitrack* ◉ *MIDI Sequencing* ◉ *Video*
- ◉ *Digital Multitrack* ◉ *DAW* ○ *Film*

Other Resources
- ○ *Professional Studios* ○ *Television Stations*
- ○ *Radio Stations* ○ *Theater Tech Dept*

Class Size Lecture	*30*	**Class Size Lab**	*3*

Assistance
- ◉ *Housing* ◉ *Scholarships* ◉ *Internships*
- ◉ *Financial Aid* ◉ *Work-study* ◉ *Job Placement*

Admission Policy	*Selective*	**Language**	*English*

Prerequisites
- ◉ *High School Diploma* ○ *SAT-ACT* ○ *Music Audition*

Industry Affiliations
- ○ *AES* ○ *ASA* ○ *NACB* ○ *NARAS* ○ *SMPTE* ○ *NAMBI*
- ○ *APRS* ○ *NAB* ○ *NAMM* ○ *SBE* ◉ *SPARS* ◉ *MEIEA*

Fulltime Faculty	*3*	**Parttime Faculty**	*10*

Faculty Awards

Program Awards

Student Awards

Research areas

New York University, Department of Music

NYU offers several options for studying music technology, recording, business, and entertainment professions through their large and active music department. The Music Business and Technology (MBT) division of the school emphasizes integration of musical and technical skills. MBT maintains close alliance with the Performance and Composition programs, encouraging collaborations that explore new relationships between experimental and traditional approaches to music. Students are also encouraged to participate in a wide-ranging internship program, placing them for a semester in a professional environment such as a record company, recording studio, or publishing house. Students may earn a Bachelor of Music in either Music Business or Music Technology, Master of Arts in Music Entertainment Professions or a Master of Music degree in Music Technology. Honors students can participate in the Stephen F. Temmer Tonmeister program.

The Stephen F. Temmer Tonmeister program is a graduate-level, highly-selective program that admits approximately five students per year. The Tonmeister studies emphasize a coordination of musical and technical skills, enabling participants to direct live concert recordings with a sensitivity to the demands of both disciplines. Admission requires an undergraduate degree in music or music technology, experience in music technology and recording, and an entrance examination which test ear training, music theory, music literature, music technology and recording. Students skills are refined during intensive workshop sessions, where participants record concerts by professional concert artists under the supervision of international recording specialists. These studies are also offered during special summer workshops.

The MBT division maintains ten recording and computer music studios, including two recording suites, four computer music laboratories, an A/V and film music editing studio, an analog synthesis studio, and two computer-based research and development facilities. MBT students also have access to the Arts and Media Laboratory, where they can gain exposure to advanced hardware platforms for computer music, graphics, animation, and multi-media. Students can realize collaborative projects with students from other NYU programs which combine arts and technology, such as the Interactive Telecommunications Project and the Tisch Film School.

The school maintains MBT Records to allow students to experience all aspects of record production within an educational setting. Students are in charge of all aspects of the organization from recording and producing to manufacturing, marketing, promoting, and administrating. The company releases one record each year. In addition to providing a unique educational experience, MBT Records provides an outlet for music from within the NYU community.

Depending on interests and abilities, students may become interns to further their studies at a variety of music industry enterprises. These internships often lead to permanent employment in the industry. Internships are available to juniors, Seniors, and graduate students. NYU students also have opportunities to study with some of the top professionals in the business through classes, independent study, and guest lectures.

Northeast Broadcasting School

Address	*142 Berkeley St*		
	Boston	*MA*	
	02116	*USA*	
Phone	*617-267-7910*	**FAX**	*617-236-7883*
Director	*Robert Matorin*	**Admission Contact**	*John Murphy*
Program Founded	*1952*	**School Type**	*Trade School*
Program Offered	*Recording Arts, Radio & Television Broadcasting*		
Degrees Offered	*Diploma*		
Program Length	*8 months*		
Estimated Tuition	*$6,800-$8,200*		
Main Emphasis	*Audio Engineering*	**Program is**	*Semi-technical*
Accreditations	*CCA*		
Number of Studios	*2*	**Is school non-profit?**	*No*
Types of Studios	○ *Acoustic Research* ○ *Electronic Music* ○ *Film/Foley* ◉ *Radio* ○ *Audio Research* ◉ *Music Recording* ◉ *Television* ○ *Video*		
Types of Recording	○ *Analog Multitrack* ◉ *MIDI Sequencing* ◉ *Video* ◉ *Digital Multitrack* ○ *DAW* ○ *Film*		
Other Resources	○ *Professional Studios* ○ *Television Stations* ○ *Radio Stations* ○ *Theater Tech Dept*		
Class Size Lecture	*10*	**Class Size Lab**	*10*
Assistance	○ *Housing* ○ *Scholarships* ◉ *Internships* ◉ *Financial Aid* ○ *Work-study* ◉ *Job Placement*		
Admission Policy	*Selective*	**Language**	*English*
Prerequisites	◉ *High School Diploma* ○ *SAT-ACT* ○ *Music Audition* *Interview*		
Industry Affiliations	○ *AES* ○ *ASA* ○ *NACB* ○ *NARAS* ○ *SMPTE* ○ *NAMBI* ○ *APRS* ○ *NAB* ○ *NAMM* ○ *SBE* ○ *SPARS* ○ *MEIEA*		
Fulltime Faculty		**Parttime Faculty**	*23*
Faculty Awards			
Program Awards			
Student Awards			
Research areas			

Northeast Broadcasting School

The Northeast Broadcasting School offers programs in Recording Arts and Radio & Television Broadcasting. The school features practical classroom instruction by experienced professional, hands-on training in radio, television, and recording studios as well as internships, personalized career counseling and lifetime job placement assistance. All students take a broad range of courses to explore many career options and to form a solid foundation of skills which can be utilized throughout their career. Students can focus on areas of special interest by choosing a program, a major, and individual elective courses.

The Recording Arts curriculum focuses on audio recording and MIDI production. The Broadcasting program allows students to select an emphasis in either radio or television. All the programs take advantage of a variety of production facilities. These include radio classrooms, each with an attached radio studio. Television students benefit from two television studios, each with its own editing system. The school's third edit suite features computer graphics and digital effects equipment. Recording Arts students train in multitrack and MIDI recording and production studios. Students also work independently in a MIDI production lab.

The Northeast Broadcasting School is located in the Back Bay neighborhood of Boston. The school offers financial aid to all qualified students and is expressly dedicated to admitting and assisting qualified students without any discrimination. Evening programs are also offered to provide access to education and career opportunities to working students.

Northeast Community College

Address	*801 E Benjamin Avenue, PO Box 469*		
	Norfolk	*NE*	
	68701	*USA*	
Phone	*402-644-0506*	**FAX**	*402-644-0560*
Director	*Timothy Miller*	**Admission Contact**	*Kris Nelson*
Program Founded	*1982*	**School Type**	*Community College*
Program Offered	*Audio and Recording Technology*		
Degrees Offered	*Associate of Applied Science*		
Program Length	*2 years*		
Estimated Tuition	*$2,100 -$2,430*		
Main Emphasis	*Music Recording*	**Program is**	*Semi-technical*
Accreditations	*North Central Association*		
Number of Studios	*2*	**Is school non-profit?**	*Yes*

Types of Studios
◉ *Acoustic Research* ◉ *Electronic Music* ◉ *Film/Foley* O *Radio*
O *Audio Research* O *Music Recording* O *Television* O *Video*

Types of Recording
◉ *Analog Multitrack* ◉ *MIDI Sequencing* ◉ *Video*
O *Digital Multitrack* ◉ *DAW* O *Film*

Other Resources
◉ *Professional Studios* ◉ *Television Stations*
O *Radio Stations* ◉ *Theater Tech Dept*

Class Size Lecture *16* **Class Size Lab** *2*

Assistance
◉ *Housing* ◉ *Scholarships* O *Internships*
◉ *Financial Aid* ◉ *Work-study* O *Job Placement*

Admission Policy *Open* **Language** *English*

Prerequisites
◉ *High School Diploma* O *SAT-ACT* O *Music Audition*

Industry Affiliations
◉ *AES* O *ASA* O *NACB* O *NARAS* O *SMPTE* O *NAMBI*
O *APRS* O *NAB* O *NAMM* O *SBE* ◉ *SPARS* O *MEIEA*

Fulltime Faculty *2* **Parttime Faculty** *1*

Faculty Awards

Program Awards

Student Awards

Research areas

Northeast Community College

Northeast Community College is a comprehensive institution with over 50 programs of study in vocational, technical, liberal arts, business, agricultural, and health fields. Many of the program provide on-the-job training through the Cooperative Internship Program. Graduates are ready for immediate employment or for transfer to a four-year college. Most Northeast students come from the college's 20-county service area as well as South Dakota and Iowa.

Students enrolled in the Audio and Recording Technology program typically work towards their Associate degree over a two-year (four-semester) period. First-year classes begin each August. Total tuition for the program is $2,100 for Nebraska residents and $2,430 for non-resident students.

The typical student in the Audio and Recording Technology program brings a strong interest in music, with some having a music performance background as well as an above average scholastic ability. While the primary thrust of the program is music recording, students are exposed to concert sound and lighting reinforcement, MIDI production, commercial production, digital and analog editing, studio design, electronics, and music theory.

Some unique characteristics of the program are small classes, individualized studio labs, and 24-hour access to lab facilities for students with sophomore standing. Student recording sessions utilize musicians from the community, campus groups, student bands, and individual class talent. Some students produce their own albums during their stay.

The program has placed graduates in studio engineering, live sound engineering, pro audio sales, audio business management, system installation, and record production.

Ohio University- Telecommunications

Address 9 South College St, Athens, OH, 45701, USA

Phone 614-593-4870 **FAX** 614-593-9184

Director Jeff Redefer **Admission Contact** Joseph Slade

Program Founded **School Type** University

Program Offered Professional Audio Production Sequence

Degrees Offered Bachelor of Science in Communications

Program Length 4 years

Estimated Tuition Inquire

Main Emphasis Broadcasting **Program is** Technical

Accreditations

Number of Studios 4 **Is school non-profit?** Yes

Types of Studios
- ○ Acoustic Research
- ◉ Electronic Music
- ○ Film/Foley
- ◉ Radio
- ○ Audio Research
- ◉ Music Recording
- ◉ Television
- ◉ Video

Types of Recording
- ◉ Analog Multitrack
- ◉ MIDI Sequencing
- ◉ Video
- ◉ Digital Multitrack
- ◉ DAW
- ◉ Film

Other Resources
- ◉ Professional Studios
- ◉ Television Stations
- ◉ Radio Stations
- ◉ Theater Tech Dept

Class Size Lecture Varies **Class Size Lab** Varies

Assistance
- ◉ Housing
- ◉ Scholarships
- ○ Internships
- ◉ Financial Aid
- ○ Work-study
- ○ Job Placement

Admission Policy Highly selective **Language** English

Prerequisites
- ◉ High School Diploma
- ◉ SAT-ACT
- ○ Music Audition

Industry Affiliations
- ◉ AES
- ○ ASA
- ○ NACB
- ○ NARAS
- ○ SMPTE
- ○ NAMBI
- ○ APRS
- ○ NAB
- ○ NAMM
- ○ SBE
- ○ SPARS
- ○ MEIEA

Fulltime Faculty 17 **Parttime Faculty** 2

Faculty Awards

Program Awards

Student Awards

Research areas

Ohio University- Telecommunications

Ohio University offers the Professional Audio Production Sequence through its School of Telecommunications in cooperation with other departments, such as Film, Music, Hearing and Speech Sciences, Business, and Theater. The four-year Bachelor of Science in Communication program is designed for students with serious academic and professional interests in audio production. The liberal arts-based curriculum covers technical theory, practical skills, and aesthetic considerations. The sequence is aimed at those interested in music recording, media/commercial production, documentary, theater/drama, and experimental media.

The Professional Audio Production Sequence consists of a core of audio production and telecommunications courses that are supplemented with classes outside the department, allowing students to tailor the program to their individual interests. Audio courses cover topics that include multitrack recording, digital audio workstations, audio post-production, basic electronics, and independent production projects. The school's recording facilities include a fully-equipped, multitrack recording studio, complete with DAW and a digital multitrack production studio with lock-to-picture. Additional studios are utilized in other departments, allowing students to take MIDI or Synclavier courses in the School of Music, work in sound design and reinforcement in the School of Theater, or engineer soundtracks at the re-recording facilities in the School of Film. Several on-campus radio and television stations also allow students to gain broadcasting and broadcast production experience.

Approximately 20 students are admitted to the program at the beginning of the academic year. Applicants must complete a core of class in telecommunications and attain a 2.67 grade point average to be qualified to transfer into the sequence.

Omega Studios' School of Recording Arts & Sciences

Address	*5909 Fishers Lane* *Rockville* *20852*	*MD* *USA*	
Phone	*301-230-9100*	**FAX**	*301-230-9103*
Director	*W. Robert Yesbek*	**Admission Contact**	*Barbara Taubersmith*
Program Founded	*1977*	**School Type**	*Trade School*
Program Offered	*Recording Engineering, Electronic Music, Live Sound, Audio for*		
Degrees Offered	*Certificate*		
Program Length	*6-38 weeks*		
Estimated Tuition	*$445-$4,040*		
Main Emphasis	*Music Recording*	**Program is**	*Semi-technical*
Accreditations	*MD Higher Ed Comm, VA*		
Number of Studios	*4*	**Is school non-profit?**	*No*
Types of Studios	○ *Acoustic Research* ◉ *Electronic Music* ○ *Film/Foley* ○ *Radio* ○ *Audio Research* ◉ *Music Recording* ○ *Television* ○ *Video*		
Types of Recording	◉ *Analog Multitrack* ◉ *MIDI Sequencing* ○ *Video* ◉ *Digital Multitrack* ◉ *DAW* ○ *Film*		
Other Resources	◉ *Professional Studios* ○ *Television Stations* ○ *Radio Stations* ○ *Theater Tech Dept*		
Class Size Lecture	*6-15*	**Class Size Lab**	*2-3*
Assistance	○ *Housing* ○ *Scholarships* ○ *Internships* ○ *Financial Aid* ○ *Work-study* ◉ *Job Placement*		
Admission Policy	*Open*	**Language**	*English*
Prerequisites	○ *High School Diploma* ○ *SAT-ACT* ○ *Music Audition*		
Industry Affiliations	◉ *AES* ○ *ASA* ○ *NACB* ◉ *NARAS* ○ *SMPTE* ○ *NAMBI* ○ *APRS* ○ *NAB* ○ *NAMM* ○ *SBE* ○ *SPARS* ○ *MEIEA*		
Fulltime Faculty	*10*	**Parttime Faculty**	*4*
Faculty Awards	*Clios, Gold/Platinum Records, Grammy*		
Program Awards			
Student Awards			
Research areas			

Omega Studios' School of Recording Arts & Sciences

The Omega Studios' School of Applied Recording Arts and Sciences offers five programs designed specifically for the aspiring recording, live sound, or audio production engineer, electronic musician/engineer, and music businessperson. Students benefit from working closely with lecturers, engineer, and instructors who are professionals competing day-to-day in the real world.

The major program offered by the school is their comprehensive Recording Engineering and Studios Techniques program. The curriculum is designed to give students a through grounding in the theory, techniques, and practices of professional multitrack music recording. Through the use of lectures, demonstrations, and hands-on experience, students prepare for entry-level positions in multitrack music recording studios. The 38-week program is offered both as a daytime or evening curriculum.

The Electronic Music Synthesizers and MIDI program begins by familiarizing students with the operation and programming of synthesizers and samplers. As they move through the program, students become familiar with popular sequencing programs, MIDI protocol, a variety of controller, and the use of a digital audio workstation. Students spend substantial group and individual project time working with the same equipment available to Omega Studios' wide client base.

The Sound Reinforcement for Live Performance and the Audio Production Techniques for Advertising share the same basic and intermediate levels of study as the Music Engineering program, but allow students to place a specific vocational focus on either sound reinforcement or radio, television, slide/film (educational/industrial) audio production or both.

The Music Business program provides an overview of the industry. Covering the history of music, recording contracts, artist management, record retailing, record budgets, and more, the program prepares students for multi-faceted career possibilities. Graduates from the program have worked in fields as diverse as record retailing, nightclub ownership and management, music production, and studio booking and sales, to name a few.

The Omega School offers both day and evening programs. New classes commence every January, April, July, and October. The equipment used for instruction is the latest state-of-the-art analog and digital recording, synthesis, and sound reinforcement equipment and is housed in professionally designed and constructed recording environments. Professional standards of operation are impressed upon students, and they are expected to maintain these standards when using the equipment as they participate in their programs.

Ontario Institute of Audio Recording Technology

Address	*502 Newbold Street*		
	London	*Ontario*	
	N6E 1K6	*Canada*	
Phone	*519-686-5010*	**FAX**	*519-686-5060*
Director	*Ken Trevenna*	**Admission Contact**	*Paul Steenhuis*
Program Founded	*1983*	**School Type**	*Trade School*
Program Offered	*Audio Recording Engineering*		
Degrees Offered	*Diploma*		
Program Length	*8 months*		
Estimated Tuition	*$5,5750*		
Main Emphasis	*Audio Engineering*	**Program is**	*Technical*
Accreditations			
Number of Studios	*3*	**Is school non-profit?**	*No*

Types of Studios
- ○ *Acoustic Research* ◉ *Electronic Music* ◉ *Film/Foley* ○ *Radio*
- ○ *Audio Research* ◉ *Music Recording* ○ *Television* ○ *Video*

Types of Recording
- ◉ *Analog Multitrack* ◉ *MIDI Sequencing* ◉ *Video*
- ○ *Digital Multitrack* ○ *DAW* ○ *Film*

Other Resources
- ◉ *Professional Studios* ○ *Television Stations*
- ○ *Radio Stations* ○ *Theater Tech Dept*

Class Size Lecture	*25*	**Class Size Lab**	*3*

Assistance
- ◉ *Housing* ○ *Scholarships* ○ *Internships*
- ◉ *Financial Aid* ○ *Work-study* ○ *Job Placement*

Admission Policy	*Selective*	**Language**	*English*

Prerequisites
- ◉ *High School Diploma* ○ *SAT-ACT* ○ *Music Audition*
- *Mature Student*

Industry Affiliations
- ◉ *AES* ○ *ASA* ○ *NACB* ○ *NARAS* ◉ *SMPTE* ○ *NAMBI*
- ◉ *APRS* ○ *NAB* ○ *NAMM* ○ *SBE* ◉ *SPARS* ○ *MEIEA*

Fulltime Faculty	*9*	**Parttime Faculty**	*0*

Faculty Awards

Program Awards

Student Awards

Research areas

Ontario Institute of Audio Recording Technology

The Ontario Institute of Audio Recording Technology was established in 1983 to fulfill the educational needs of those with career aspirations in the audio recording and post-production industries. All courses begin at an introductory level and accelerate rapidly, combining to provide students with knowledge and practical skills necessary to enter into all aspects of the music recording industry. A few of the subject areas covered in the school's intensive eight-month program include management, music production, MIDI, live sound, video and acoustics, and the use of tape recorders, mixing consoles, monitoring systems, and signal processors.

Designed in close liaison with OIART's advisory committee and the music/audio recording industry, the program emphasizes practical training in the school's three professional multitrack studios and numerous lab stations. Though the program stresses the technical recording aspects of the music business, students are reminded that they are dealing with a creative industry and are encouraged to engage in performance, composition, and songwriting.

In order to keep the atmosphere of the program both informal and personable, OIART limits enrollment to approximately 50 students per year. The nine very accessible, full-time staff members are industry professionals specialized in their own fields. Each course is implemented by the instructor who designed and developed it.The staff place great emphasis on professionalism and share over 100 years of combined experience in the industry.

Oracle Recording Studio

Address *PO Box 464188*
Lawrenceville *GA*
30246 *USA*

Phone *404-921-7941* **FAX**

Director *Gene Smith* **Admission Contact** *Gene Smith*

Program Founded *1991* **School Type** *Workshop*

Program Offered *Beginning Recording Techniques*

Degrees Offered *None*

Program Length *10 hours*

Estimated Tuition *$200*

Main Emphasis *Music Recording* **Program is** *Semi-technical*

Accreditations *None*

Number of Studios *1* **Is school non-profit?** *No*

Types of Studios
○ *Acoustic Research* ○ *Electronic Music* ○ *Film/Foley* ○ *Radio*
○ *Audio Research* ◉ *Music Recording* ○ *Television* ○ *Video*

Types of Recording
◉ *Analog Multitrack* ○ *MIDI Sequencing* ○ *Video*
○ *Digital Multitrack* ○ *DAW* ○ *Film*

Other Resources
◉ *Professional Studios* ○ *Television Stations*
○ *Radio Stations* ○ *Theater Tech Dept*

Class Size Lecture *4* **Class Size Lab** *4*

Assistance
○ *Housing* ○ *Scholarships* ○ *Internships*
○ *Financial Aid* ○ *Work-study* ○ *Job Placement*

Admission Policy *Open* **Language** *English*

Prerequisites ○ *High School Diploma* ○ *SAT-ACT* ○ *Music Audition*

Industry Affiliations
◉ *AES* ○ *ASA* ○ *NACB* ○ *NARAS* ○ *SMPTE* ○ *NAMBI*
○ *APRS* ○ *NAB* ○ *NAMM* ○ *SBE* ○ *SPARS* ○ *MEIEA*

Fulltime Faculty *1* **Parttime Faculty** *0*

Faculty Awards

Program Awards

Student Awards

Research areas

Oracle Recording Studio

Oracle Recording Studio offers a basic Studio Workshop. The Workshop consists of basic audio, console signal flow, vocal recording, signal processing, compressors and limiters, multitrack recording, microphone techniques, MIDI and sequencing, digital audio, mixing, and more. The program is customized to the student's interests. The workshop is aimed at those interested in recording their own projects or those in the process putting together a home studio. The Workshop provides plenty of hands-on experience. The course is designed to teach by association with practical exercises. Students can also take advantage of access to over 300 lines of professional audio and video equipment offered at exceptional discounts through the program.

Peabody Conservatory of the John Hopkins University

Address *1 East Mt. Vernon Pl.*
Baltimore *MD*
21202 *USA*

Phone *410-659-8136* **FAX** *410-659-8168*

Director *Alan Kefauver* **Admission Contact** *David Lane*

Program Founded *1983* **School Type** *University*

Program Offered *Recording Arts and Sciences, Electronic and Computer Music*

Degrees Offered *Bachelor of Music, Master of Music*

Program Length *4 years, 2 years*

Estimated Tuition *$15,000 per year*

Main Emphasis *Audio Engineering* **Program is** *Semi-technical*

Accreditations *NASM, ABET, Middle States*

Number of Studios *3* **Is school non-profit?** *Yes*

Types of Studios ○ *Acoustic Research* ◉ *Electronic Music* ○ *Film/Foley* ○ *Radio*
○ *Audio Research* ◉ *Music Recording* ○ *Television* ○ *Video*

Types of Recording ◉ *Analog Multitrack* ◉ *MIDI Sequencing* ◉ *Video*
◉ *Digital Multitrack* ◉ *DAW* ○ *Film*

Other Resources ◉ *Professional Studios* ○ *Television Stations*
◉ *Radio Stations* ○ *Theater Tech Dept*

Class Size Lecture *8* **Class Size Lab** *6*

Assistance ◉ *Housing* ◉ *Scholarships* ◉ *Internships*
◉ *Financial Aid* ◉ *Work-study* ◉ *Job Placement*

Admission Policy *Highly selective* **Language** *English*

Prerequisites ◉ *High School Diploma* ◉ *SAT-ACT* ◉ *Music Audition*
Interview

Industry Affiliations ◉ *AES* ◉ *ASA* ○ *NACB* ◉ *NARAS* ◉ *SMPTE* ○ *NAMBI*
○ *APRS* ◉ *NAB* ○ *NAMM* ○ *SBE* ○ *SPARS* ○ *MEIEA*

Fulltime Faculty *76* **Parttime Faculty** *67*

Faculty Awards

Program Awards *TEC Award Nominee*

Student Awards

Research areas

Peabody Conservatory of the John Hopkins University

The Peabody Conservatory of John Hopkins University offers a five-year double-major Recording Arts and Sciences program that combines the elements of a performance major with the technological resources of the Peabody Recording Studios and the John Hopkins University laboratories of the G.W.C Whiting School of Engineering, nationally-recognized for its reputation in electrical engineering and computer science. Work in the program includes extensive practical experience that culminates with 5th-year internships with local radio, television, and recording companies.

In addition to regular lab sessions, students participate in recording a wide variety of music, ranging from jazz-rock to grand opera, music for video and film, and more. The abilities of the recording studios to broadcast events live over satellite and to other divisions of the University further enhance the educational program for the students. Students receive a minimum of 2000 hours of hands-on experience in the studios, and all recording classes are held in the studios. All students are employed by the studios and are paid by work-study or student employee funds. This is possible due to the program's limited enrollment.

Entry into the program is highly competitive and prospective students must meet the entrance requirements of the Peabody Conservatory, including a music audition in the candidate's major performance area, along with the mathematics requirement of the John Hopkins University. A maximum of 10 students are accepted into the program each year.

The school's recording studio complex was completed in 1983 and is continually updated to reflect the current state-of-the-art. The facility is complete with digital and analog multitrack automated equipment and is linked to all the performance halls. There are several smaller production facilities. A separate Computer Music and MIDI studio, part of the school's Master degree program in Computer Music, is available for students to take courses in advanced MIDI and electronic composition techniques.

In 1991, Peabody Conservatory introduced a Master of Music degree in Electronic and Computer Music. The two-year program has three areas of concentration: Composition, Performance/Concert Production, and Research/Music Technology. The Composition track allows special concentration in composing music with electronic and computer music systems. The Performance/Concert Production track allows students to gain skills and sensibilities necessary to become concert performers with new technology, as well as to produce entire multi-media concerts. The Research/Technology track is designed for students pursuing musically-related research or developing new music technologies. All of these Master of Music programs require Bachelor of Music degrees and competency in high-level computer programming languages in addition to other prerequisites.

Pennsylvania State University

Address	*220 Special Services Building* *University Park* *PA* *16802* *USA*		
Phone	*814-863-2911*	**FAX**	*814-863-2574*
Director	*Peter Kiefer*	**Admission Contact**	*Peter Kiefer*
Program Founded	*1980*	**School Type**	*University*
Program Offered	*Sound and Recording Workshop*		
Degrees Offered	*None*		
Program Length	*4 days*		
Estimated Tuition	*$185- $275*		
Main Emphasis	*Music Recording*	**Program is**	*Semi-technical*
Accreditations			
Number of Studios	*1*	**Is school non-profit?**	*Yes*
Types of Studios	O *Acoustic Research* O *Electronic Music* O *Film/Foley* O *Radio* O *Audio Research* O *Music Recording* ◉ *Television* O *Video*		
Types of Recording	◉ *Analog Multitrack* O *MIDI Sequencing* ◉ *Video* ◉ *Digital Multitrack* O *DAW* O *Film*		
Other Resources	◉ *Professional Studios* ◉ *Television Stations* O *Radio Stations* O *Theater Tech Dept*		
Class Size Lecture	*25*	**Class Size Lab**	*25*
Assistance	◉ *Housing* O *Scholarships* O *Internships* O *Financial Aid* O *Work-study* O *Job Placement*		
Admission Policy	*Open*	**Language**	*English*
Prerequisites	O *High School Diploma* O *SAT-ACT* O *Music Audition*		
Industry Affiliations	◉ *AES* O *ASA* O *NACB* ◉ *NARAS* O *SMPTE* O *NAMBI* O *APRS* ◉ *NAB* O *NAMM* O *SBE* O *SPARS* O *MEIEA*		
Fulltime Faculty	*2*	**Parttime Faculty**	
Faculty Awards			
Program Awards			
Student Awards			
Research areas			

Pennsylvania State University

During the summer, Penn State offers a Sound and Recording Workshop for music teachers, directors, young people, and others seeking experience in and knowledge of sound reinforcement and recording. Participants receive practical information that will help them to select and operate equipment while achieving optimum performance. Workshop sessions are held in the Eisenhower Auditorium on Penn State's University Park Campus.

Topics covered by the workshop include: microphones, mixers, amplifiers, speakers, recorders, sound reinforcement, concert setup techniques, and hands-on experience in live concert settings. Undergraduate or graduate credit is available for students. High school students also receive college credit and can sign up as undergraduates for the workshop. Accommodations for participants are available in University residence halls.

Clinicians Peter Kiefer and Kerry Trout are both active sound engineers with considerable experience in concert and studio recording and sound reinforcement.

Pennsylvania State University

Address	*Graduate Program in Acoustics, Applied Science Bldg* *University Park* *PA* *16802* *USA*
Phone	*814-865-6364* **FAX** *814-865-3119*
Director	*Jiri Tichy* **Admission Contact** *Barbara Crocken*
Program Founded	*1965* **School Type** *University*
Program Offered	*Graduate Acoustics*
Degrees Offered	*Masters, Ph.D.*
Program Length	*Varies*
Estimated Tuition	*Inquire*
Main Emphasis	*Acoustic Design* **Program is** *Technical*
Accreditations	
Number of Studios	*12* **Is school non-profit?** *Yes*
Types of Studios	● *Acoustic Research* ○ *Electronic Music* ○ *Film/Foley* ○ *Radio* ○ *Audio Research* ○ *Music Recording* ○ *Television* ○ *Video*
Types of Recording	○ *Analog Multitrack* ○ *MIDI Sequencing* ○ *Video* ○ *Digital Multitrack* ○ *DAW* ○ *Film*
Other Resources	○ *Professional Studios* ○ *Television Stations* ○ *Radio Stations* ○ *Theater Tech Dept* *Acoustic Research Labs*
Class Size Lecture	*varies* **Class Size Lab** *varies*
Assistance	● *Housing* ● *Scholarships* ○ *Internships* ● *Financial Aid* ● *Work-study* ○ *Job Placement*
Admission Policy	*Highly selective* **Language** *English*
Prerequisites	● *High School Diploma* ○ *SAT-ACT* ○ *Music Audition* *BS in Engineering*
Industry Affiliations	● *AES* ● *ASA* ○ *NACB* ○ *NARAS* ○ *SMPTE* ○ *NAMBI* ○ *APRS* ○ *NAB* ○ *NAMM* ○ *SBE* ○ *SPARS* ○ *MEIEA*
Fulltime Faculty	*6* **Parttime Faculty** *30*
Faculty Awards	
Program Awards	
Student Awards	
Research areas	

Pennsylvania State University

The Graduate Program in Acoustics at Penn State has grown from under 10 students when it was founded in 1965 to over 120 students in 1992. At Penn State, an "acoustics community", consisting of several colleges and departments, is unified by the Graduate Program which offers M.Eng., M.S., and Ph.D. degrees in Acoustics. Penn State offers a interdisciplinary program where students are encouraged to build a broad acoustics background while specializing on their thesis topic. Students are assisted by advisors who help frame a curriculum which meets the core requirements of the program as well as the students' goals and needs in a particular area of acoustics.

The acoustics faculty are more than willing to discuss their work with students to help develop better understanding of acoustics. More than 36 faculty members associated with the program work in general areas of underwater acoustics and transducers, noise control, active control of sound and vibration, structural acoustics, acoustic signal processing, and turbulence and fluid dynamics. This is augmented by prominent acoustics researchers from around the world who visit the program and give seminars. Faculty from the Applied Research Laboratory (ARL), a leading acoustics research facility, support the Acoustics Program with thesis advising, research assistantships, and technical support for students doing research related to ongoing projects at ARL.

Other facilities available for student research include: the ARL Anechoic Chamber, the Noise Control Lab, the Acousto-optic Imaging Lab, the Acoustic Intensity Lab, the Acoustics/Physics Lab, the Shock and Vibration Lab, the High-Intensity Acoustics Lab, the Dynamic Materials Lab, the Dynamic Response Lab, the Human Research Lab, the Medical Ultrasonic Lab, and the Ultrasonic Diagnostic Lab.

Because of its reputation as a center of acoustics research, Penn State has hosted numerous national and international conventions, including the Institute of Noise Control Engineering's Noise-Con Conference in 1987 and the Acoustical Society of America's spring 1990 meeting.

Recording Arts Program of Canada

Address	*28 Valrose Drive* *Stoney Creek* *Ontario* *L8E 3T4* *Canada*
Phone	*416-662-2666* **FAX** *416-662-2666*
Director	*John Keca* **Admission Contact** *John Keca*
Program Founded	*1982* **School Type** *Trade School*
Program Offered	*Audio Engineering/Production*
Degrees Offered	*Diploma*
Program Length	*1 year*
Estimated Tuition	*$4,000*
Main Emphasis	*Audio Engineering* **Program is** *Technical*
Accreditations	*US Dept of Education*
Number of Studios	*3* **Is school non-profit?** *No*
Types of Studios	○ *Acoustic Research* ● *Electronic Music* ○ *Film/Foley* ○ *Radio* ○ *Audio Research* ● *Music Recording* ○ *Television* ○ *Video*
Types of Recording	● *Analog Multitrack* ● *MIDI Sequencing* ○ *Video* ○ *Digital Multitrack* ● *DAW* ○ *Film*
Other Resources	● *Professional Studios* ○ *Television Stations* ○ *Radio Stations* ○ *Theater Tech Dept*
Class Size Lecture	*10* **Class Size Lab** *2*
Assistance	● *Housing* ○ *Scholarships* ○ *Internships* ● *Financial Aid* ○ *Work-study* ○ *Job Placement*
Admission Policy	*Open* **Language** *English*
Prerequisites	● *High School Diploma* ○ *SAT-ACT* ○ *Music Audition*
Industry Affiliations	○ *AES* ○ *ASA* ○ *NACB* ○ *NARAS* ○ *SMPTE* ○ *NAMBI* ○ *APRS* ○ *NAB* ○ *NAMM* ○ *SBE* ○ *SPARS* ○ *MEIEA*
Fulltime Faculty	*4* **Parttime Faculty** *4*
Faculty Awards	
Program Awards	
Student Awards	
Research areas	

Recording Arts Program of Canada

The Recording Arts Program (RAP) of Canada offers a one-year program in Audio Engineering and Production. The focus of the intensive program is to explore the fundamentals of the recording process and current technology. The program maintains multitrack studios, audio post-production suites, digital audio workstations, and comprehensive MIDI systems. Engineering and production classes are limited to ten students, and over 60% of the instruction is taught directly in the studios.

The program's fundamental intention is to make students conversant with contemporary technology found in all production arenas. The primary areas of instruction are: recording engineering, audio post-production, MIDI, and digital audio recording and editing. Each component is given an equal amount of attention with continual overlap or interaction with the other components.

At the Stoney Creek facility, 40 minutes south of Toronto, the program benefits from two on-site recording studios as well as opportunities to utilize Aztec Studios, an independent, active commercial recording facility. The owner of Aztec Studios, Curtis Lambert, is also an instructor in the program.

The program commences every September and March. The school offers career day seminars to allow prospective students to observe the various facilities in order to illustrate the types of projects students can explore. The program also has a quarterly newsletter as well as a short video that provides a closer look at RAP.

Recording Arts Program of Canada

Address *34 Chemin des Ormes*
Ste-Anne-des-Lacs *Quebec*
J0R 1B0 *Canada*

Phone *514-224-8363* **FAX** *416-662-2666*

Director *John Keca* **Admission Contact** *John Keca*

Program Founded *1982* **School Type** *Trade School*

Program Offered *Audio Engineering/Production*

Degrees Offered *Diploma*

Program Length *1 academic year*

Estimated Tuition *$4,000*

Main Emphasis *Audio Engineering* **Program is** *Technical*

Accreditations *US Dept. of Education*

Number of Studios *1* **Is school non-profit?** *No*

Types of Studios ○ *Acoustic Research* ◉ *Electronic Music* ○ *Film/Foley* ○ *Radio*
○ *Audio Research* ◉ *Music Recording* ○ *Television* ○ *Video*

Types of Recording ◉ *Analog Multitrack* ◉ *MIDI Sequencing* ○ *Video*
○ *Digital Multitrack* ◉ *DAW* ○ *Film*

Other Resources ◉ *Professional Studios* ○ *Television Stations*
○ *Radio Stations* ○ *Theater Tech Dept*

Class Size Lecture *10* **Class Size Lab** *2*

Assistance ◉ *Housing* ○ *Scholarships* ○ *Internships*
◉ *Financial Aid* ○ *Work-study* ○ *Job Placement*

Admission Policy *Open* **Language** *English*

Prerequisites ◉ *High School Diploma* ○ *SAT-ACT* ○ *Music Audition*

Industry Affiliations ○ *AES* ○ *ASA* ○ *NACB* ○ *NARAS* ○ *SMPTE* ○ *NAMBI*
○ *APRS* ○ *NAB* ○ *NAMM* ○ *SBE* ○ *SPARS* ○ *MEIEA*

Fulltime Faculty *4* **Parttime Faculty** *4*

Faculty Awards

Program Awards

Student Awards

Research areas

Recording Arts Program of Canada

The Recording Arts Program (RAP) of Canada offers a one-year program in Audio Engineering and Production. The focus of the intensive program is to explore the fundamentals of the recording process and current technology. The program maintains multitrack studios, audio post-production suites, digital audio workstations, and comprehensive MIDI systems. Engineering and production classes are limited to ten students, and over 60% of the instruction is taught directly in the studios.

The program's fundamental intention is to make students conversant with contemporary technology found in all production arenas. The primary areas of instruction are: recording engineering, audio post-production, MIDI, and digital audio recording and editing. Each component is given an equal amount of attention with continual overlap or interaction with the other components.

At the Quebec facility, located in the Laurentian Mountains 40 minutes north of Montreal, consists of a three room studio that is interface to accommodate recording in up to seven other rooms in the building. The studio is equipped with classic analog recording hardware as well as modern music technology.

The program commences every September and March. The school offers career day seminars to allow prospective students to observe the various facilities in order to illustrate the types of projects students can explore. The program also has a quarterly newsletter as well as a short video that provides a closer look at RAP.

Recording Institute of Detroit

Address *14611 E Nine Mile Road*
Eastpoint *MI*
48021 *USA*

Phone *313-779-1380* **FAX** *313-772-4320*

Director *Robert Dennis* **Admission Contact**

Program Founded *1975* **School Type** *Trade School*

Program Offered *Multitrack Recording Techniques*

Degrees Offered *Certificate*

Program Length *39 weeks*

Estimated Tuition *$3,195*

Main Emphasis *Music Recording* **Program is** *Technical*

Accreditations *Michigan Dept of Education*

Number of Studios *3* **Is school non-profit?** *No*

Types of Studios ○ *Acoustic Research* ○ *Electronic Music* ○ *Film/Foley* ○ *Radio*
○ *Audio Research* ◉ *Music Recording* ○ *Television* ○ *Video*

Types of Recording ◉ *Analog Multitrack* ○ *MIDI Sequencing* ○ *Video*
○ *Digital Multitrack* ○ *DAW* ○ *Film*

Other Resources ○ *Professional Studios* ○ *Television Stations*
○ *Radio Stations* ○ *Theater Tech Dept*

Class Size Lecture *14* **Class Size Lab** *8*

Assistance ○ *Housing* ○ *Scholarships* ◉ *Internships*
○ *Financial Aid* ○ *Work-study* ◉ *Job Placement*

Admission Policy *Open* **Language** *English*

Prerequisites ○ *High School Diploma* ○ *SAT-ACT* ○ *Music Audition*
Scholastic Level Exam

Industry Affiliations ○ *AES* ○ *ASA* ○ *NACB* ○ *NARAS* ○ *SMPTE* ○ *NAMBI*
○ *APRS* ○ *NAB* ○ *NAMM* ○ *SBE* ○ *SPARS* ○ *MEIEA*

Fulltime Faculty *3* **Parttime Faculty** *3*

Faculty Awards

Program Awards

Student Awards

Research areas

Recording Institute of Detroit

The Recording Institute of Detroit is organized for those interested in the latest recording techniques either professionally, as a hobby, or as part of a related industry such as record production, live-sound mixing, or television/radio broadcasting and production. The school offers a 376-hour Audio Techniques training program that can be completed in 39 weeks by attending 9-10 hours per week. The program provides training for entry level positions such as recording engineer, second recording engineer, and sound/recording technician.

The Recording Institute of Detroit's program is divided into three 13-week quarters. The first quarter consists of the Basic Recording Course and the Applied Music Theory Course. The second is devoted to the Advanced Recording Course where students apply their knowledge to a minimum of four recording projects and to a final demonstration of all the recording techniques they have learned. During the third quarter, students finish the Advanced Recording Course by completing lab time and assignments by functioning as a second recording engineer on sessions. An internship of 100 hours follows. Students can intern in recording or related fields. The program can be taken as individual classes or as one program.

The program benefits from three professional recording studios as well as two learning stations. The school also has a bookstore that maintains a stock of alternative texts and recording supplies related to the program. Orientations and open houses are regularly held to familiarize students to the program and its facilities.

Recording Workshop

Address	*455 Massieville Road* *Chillicothe* *OH* *45601* *USA*
Phone	*800-848-9900* **FAX** *614-663-2427*
Director	*Jim Rosebrook* **Admission Contact** *Paul Pollard*
Program Founded	*1971* **School Type** *Workshop*
Program Offered	*Recording Engineering and Music Production*
Degrees Offered	*Certificate*
Program Length	*5 weeks*
Estimated Tuition	*$2,090*
Main Emphasis	*Music Recording* **Program is** *Semi-technical*
Accreditations	*Ohio State Board of Schools*
Number of Studios	*6* **Is school non-profit?** *No*
Types of Studios	○ *Acoustic Research* ○ *Electronic Music* ○ *Film/Foley* ○ *Radio* ○ *Audio Research* ◉ *Music Recording* ○ *Television* ○ *Video*
Types of Recording	◉ *Analog Multitrack* ◉ *MIDI Sequencing* ○ *Video* ◉ *Digital Multitrack* ◉ *DAW* ○ *Film*
Other Resources	○ *Professional Studios* ○ *Television Stations* ○ *Radio Stations* ○ *Theater Tech Dept* *Capital University*
Class Size Lecture	*36* **Class Size Lab** *6*
Assistance	◉ *Housing* ○ *Scholarships* ◉ *Internships* ◉ *Financial Aid* ○ *Work-study* ◉ *Job Placement*
Admission Policy	*Open* **Language** *English*
Prerequisites	◉ *High School Diploma* ○ *SAT-ACT* ○ *Music Audition*
Industry Affiliations	◉ *AES* ○ *ASA* ○ *NACB* ○ *NARAS* ○ *SMPTE* ○ *NAMBI* ○ *APRS* ○ *NAB* ○ *NAMM* ○ *SBE* ◉ *SPARS* ○ *MEIEA*
Fulltime Faculty	*14* **Parttime Faculty** *4*
Faculty Awards	*db Award, Silver Mic Award*
Program Awards	*TEC Nominee*
Student Awards	*Gold & Platinum Albums, Ampex Golden Reel*
Research areas	

Recording Workshop

The Recording Workshop, now in its third decade, is one of the largest educational facilities of its kind, offering a variety of training opportunities for those seeking careers in the field of audio engineering. The main program, Recording Engineering and Music Production, is a 5-week, 200 hour, hands-on approach to learning many aspects of audio engineering. The curriculum offers training in recording album and demo projects, music production, mixing, editing, mastering, media production, and sound effect recording for film and video. The Recording Workshop offers an alternative to traditional university education, yet can supplement many college audio programs.

The Recording Workshop is comprised of six studios. Workshop students learn on the type of recording equipment they are most likely to encounter in the job market. Students also learn the latest audio technology. The education offers training in contemporary analog multitrack recording in addition to digital multitrack recording. Macintosh-based direct-to-disk recording and digital editing are also part of the curriculum.

The studios consist of a multitrack media production studio with a complete MIDI setup, four multitrack recording studios- two with automated consoles, an analog editing suite, and a digital editing and mastering suite. The Recording Workshop continually upgrades facilities to reflect the latest technology.

Workshop students need no previous experience. During the five-week period, classes are held five days a week, eight hours a day. The daily classes consist of five hours of in-studio, hands-on training and three hours of supplemental lectures. The in-house classes are comprised of only three to six students per instructor. A typical lecture consists of thirty to thirty-six students.

The Recording Workshop also offers three optional programs. The one-week, 40-hour Studio Maintenance and Troubleshooting program teaches studio maintenance, tape machine alignment, and basic repair. The one-week, 60-hour Advanced Recording Engineering and Music Production program features recording and mixing in smaller classes without any lectures. The one-week, 35-hour New Tech Production program is exclusively dedicated to learning computer-based audio systems as well as their application to the audio job market.

The Recording Workshop offers classes seven different times a year. They also offer low-cost, on-campus housing, financial assistance, and job placement.

San Francisco State University Extended Education

Address	*425 Market Street*		
	San Francisco	*CA*	
	94105	*USA*	
Phone	*415-904-7720*	**FAX**	*415-904-7760*
Director	*Mary Pieratt*	**Admission Contact**	*Mary Pieratt*
Program Founded	*1980*	**School Type**	*University*
Program Offered	*Music/Recording Industry*		
Degrees Offered	*Certificate*		
Program Length	*1 year*		
Estimated Tuition	*$2,500*		
Main Emphasis	*Audio Engineering*	**Program is**	*Semi-technical*
Accreditations			
Number of Studios	*2-3*	**Is school non-profit?**	*Yes*
Types of Studios	○ *Acoustic Research* ○ *Electronic Music* ○ *Film/Foley* ○ *Radio* ○ *Audio Research* ◉ *Music Recording* ○ *Television* ○ *Video*		
Types of Recording	○ *Analog Multitrack* ○ *MIDI Sequencing* ○ *Video* ◉ *Digital Multitrack* ○ *DAW* ○ *Film*		
Other Resources	○ *Professional Studios* ○ *Television Stations* ○ *Radio Stations* ○ *Theater Tech Dept*		
Class Size Lecture	*25*	**Class Size Lab**	*12*
Assistance	○ *Housing* ○ *Scholarships* ○ *Internships* ○ *Financial Aid* ○ *Work-study* ○ *Job Placement*		
Admission Policy	*Selective*	**Language**	*English*
Prerequisites	◉ *High School Diploma* ○ *SAT-ACT* ○ *Music Audition* *56 credit hours of study*		
Industry Affiliations	◉ *AES* ○ *ASA* ○ *NACB* ◉ *NARAS* ◉ *SMPTE* ○ *NAMBI* ○ *APRS* ○ *NAB* ◉ *NAMM* ○ *SBE* ○ *SPARS* ○ *MEIEA*		
Fulltime Faculty	*0*	**Parttime Faculty**	*15*
Faculty Awards			
Program Awards			
Student Awards			
Research areas			

San Francisco State University Extended Education

San Francisco State University offers an academic program for those who want to gain experience and professional education in the music recording industry through a structured course of study. The Music/Recording Industry Certificate program is designed both for the student interested in an overall basic education in the business of music/recording and for the professional who is seeking to broaden an area of expertise. Coursework in the program may be applied toward the Certificate or individual classes may be taken to learn or update skills. Students may choose to pursue a technical audio engineering or a business emphasis.

The Music/Recording Industry Certificate program was developed in consultation with Bay Area music industry institutions, organizations, and individuals in an effort to align the educational goals of the program with the needs and expectations of the music and recording community. Program faculty are industry professionals from the community. In addition, the program maintains an Academic Program Committee comprised of University chairs, administration, and faculty who assure the academic standards of the program.

The San Francisco State University campus houses a complete, state-of-the-art recording studio for MRI recording workshops. Workshops are also offered in off-campus, professional recording studios in the Bay Area. Students have access to University facilities including library, gymnasium, and Student Union. Courses in the program are held on campus or in the University's Downtown San Francisco Center.

Applicants are expected to have completed a minimum of 56 college semester units/credits at the time they apply. Those not meeting this requirement may be accepted on a probationary basis. All courses in the MRI program may be used toward a Bachelor of Arts degree at all California State University institutions. San Francisco State University offers a Bachelor of Arts in Broadcast Communication Arts with an Audio Production emphasis as a four-year program. They also offer a Master of Arts program to accommodate graduate students interested in audio production, aesthetics, and criticism.

San Jose State University, Electro-Acoustic Studios

Address *One Washington square*
San Jose *CA*
95192-0095 *USA*

Phone *408-924-4773* **FAX** *408-924-4773*

Director *Allen Strange* **Admission Contact** *Allen Strange*

Program Founded *1989* **School Type** *University*

Program Offered *Electro-Acoustics*

Degrees Offered *Bachelor of Arts or Music, Master of Arts*

Program Length *4 years, 2 years*

Estimated Tuition *$5,600*

Main Emphasis *Electronic Music* **Program is** *Semi-technical*

Accreditations *NASM*

Number of Studios *4* **Is school non-profit?** *Yes*

Types of Studios ○ *Acoustic Research* ◉ *Electronic Music* ○ *Film/Foley* ○ *Radio*
○ *Audio Research* ◉ *Music Recording* ○ *Television* ○ *Video*

Types of Recording ◉ *Analog Multitrack* ○ *MIDI Sequencing* ○ *Video*
○ *Digital Multitrack* ○ *DAW* ○ *Film*

Other Resources ◉ *Professional Studios* ◉ *Television Stations*
◉ *Radio Stations* ◉ *Theater Tech Dept*

Class Size Lecture *10* **Class Size Lab** *10*

Assistance ○ *Housing* ◉ *Scholarships* ○ *Internships*
○ *Financial Aid* ○ *Work-study* ○ *Job Placement*

Admission Policy *Highly selective* **Language** *English*

Prerequisites ◉ *High School Diploma* ○ *SAT-ACT* ◉ *Music Audition*

Industry Affiliations ◉ *AES* ○ *ASA* ○ *NACB* ◉ *NARAS* ○ *SMPTE* ○ *NAMBI*
○ *APRS* ○ *NAB* ○ *NAMM* ○ *SBE* ○ *SPARS* ○ *MEIEA*

Fulltime Faculty *30* **Parttime Faculty** *20*

Faculty Awards

Program Awards

Student Awards

Research areas

San Jose State University, Electro-Acoustic Studios

San Jose State's Electro-Acoustic Studio was established in 1972 and hosted the International Computer Music Conference in 1992. The facility consists of four major installations dedicated to composition, production, and research in technical music media. The studios support a Bachelor of Arts program and graduate work in Electro-Acoustics consisting of studies in electronic/computer music, recording arts, and music production. In order to accommodate an extensive range of student interests, instruction is available in a wide range of esthetic vocabularies ranging from experimental and extend musical styles to jazz and popular music. Special emphasis is placed on technical applications in live performance, improvisation and personal software development.

Courses in the program may be taken by any qualified matriculated student in the University. Some courses are also available to the community through the SJSU Open University Program. The studios are active across the campus providing various services for other disciplines such as Theater Arts, Dance, Radio, Television, and Advertising. The Electronic/Computer Music studios are also associated with the CADRE Institute (Computers in Art, Research, Design, and Education), and its current projects are based on software development for computer animation and music.

A variety of student projects are produced in the studios; ranging from commercial spots, theater scores, commercial tunes, and large scale experimental and concert works. The Electro-Acoustic program faculty, staff, and resident guest composers have produced numerous works for a variety of media. The creative work of students, faculty, and guest artists are featured in two annual student concerts and an American Music Week festival. Compositions created in the studios have received international recognition.

In addition to the prescribed curricular responsibilities, the studios actively engage in ongoing compositional and research projects. Specific areas of research include language development, alternate MIDI controller, and digital synthesis. The program is supported by several multitrack recording and electronic music studio. A Music Minor in Electro-Acoustics is also available.

School of Audio Engineering- Adelaide

Address *18-20 Deeds Road*
Adelaide *South Australia*
5038 *Australia*

Phone *08-376-0991* **FAX** *08-238-8028*

Director *Tom Misner* **Admission Contact** *Michael Davidson*

Program Founded *1976* **School Type** *Trade School*

Program Offered *Audio Engineer, Studio Assistant, Live Sound*

Degrees Offered *Diploma, Certificate, Certificate*

Program Length *15 months, 6 months, 6 months*

Estimated Tuition *$300 per month*

Main Emphasis *Music Recording* **Program is** *Technical*

Accreditations *Dept of Ed*

Number of Studios *2* **Is school non-profit?** *No*

Types of Studios
○ *Acoustic Research* ○ *Electronic Music* ○ *Film/Foley* ○ *Radio*
○ *Audio Research* ◉ *Music Recording* ○ *Television* ○ *Video*

Types of Recording
◉ *Analog Multitrack* ◉ *MIDI Sequencing* ○ *Video*
○ *Digital Multitrack* ○ *DAW* ○ *Film*

Other Resources
◉ *Professional Studios* ○ *Television Stations*
○ *Radio Stations* ○ *Theater Tech Dept*

Class Size Lecture *20-23* **Class Size Lab** *1*

Assistance
○ *Housing* ◉ *Scholarships* ◉ *Internships*
○ *Financial Aid* ○ *Work-study* ◉ *Job Placement*

Admission Policy *Open* **Language** *English*

Prerequisites ○ *High School Diploma* ○ *SAT-ACT* ◉ *Music Audition*
16 years of age, hearing & pretest

Industry Affiliations
◉ *AES* ○ *ASA* ○ *NACB* ○ *NARAS* ○ *SMPTE* ○ *NAMBI*
○ *APRS* ○ *NAB* ○ *NAMM* ○ *SBE* ○ *SPARS* ○ *MEIEA*

Fulltime Faculty *2* **Parttime Faculty** *3*

Faculty Awards

Program Awards

Student Awards

Research areas

School of Audio Engineering- Adelaide

The School of Audio Engineering was founded by Tom Misner in 1976 and has since grown to a network of 16 schools in Europe, Asia, and Australia. The school's program in Paris is the most recent, and plans are underway for a new school in Moscow, Russia. It is the largest audio training program in the world with a total student population of over 4400 worldwide. The basic philosophy of SAE training is to give students a sold theoretical background and individual hands-on practical studio time. The school offers several programs and is constantly updating their curriculum to allow for the latest developments in audio technologies and techniques.

The main concept behind all SAE courses is individual practical studio time, allowing students to learn how to operate a professional recording studio. Theoretical lectures are designed to allow a natural flow of information to compliment this practical experience. The programs are focused to avoid confusing novices to the industry.

The Audio Engineer program is an intensive training program which gives students solid grounding in all aspects of studio and live sound work. A comprehensive list of major topics are covered following a natural learning curve. Upon completion of this program, students can opt to take SAE's three-month Production Course to further study advanced audio topics.

The Tonmeister Degree program requires applicants to have a formal music education, sight-reading skills, and a SAE Audio Engineer Diploma of equivalent course. This highly selective course is primarily suited to record producers/engineers interested in working with classical music recording.

The Studio Assistant course is for musicians who wish to further their knowledge of recording and those interested in home-studio production. It provides entry-level courses along with some practical studio time.

In addition to these programs, SAE offers seminars such as Music Business, Hard Disk Studio, Recording Studio Design and Construction, and MIDI. The seminars offered vary depending on the SAE facility presenting the program.

SAE produces a regular newsletter that is distributed to recording studios, broadcasting facilities, and others who might employ the school's graduates. Student CDs are also produced to give students experience and actual album credits. All prospective students can visit SAE's facilities by appointment or on a number of open houses the school arranges every year.

School of Audio Engineering- Amsterdam

Address	*Vondelstraat 13* *Amsterdam* *1054 GC*	*Netherlands*	
Phone	*31-20-689- 4189*	**FAX**	*31-20-689-4324*
Director	*Tom Misner*	**Admission Contact**	*Guy Nicholson*
Program Founded	*1976*	**School Type**	*Trade School*
Program Offered	*Audio Engineering, Tonmeister, Live Sound*		
Degrees Offered	*Diploma, Degree, Certificate*		
Program Length	*18 months, 12 months, 3 months*		
Estimated Tuition	*$300 per month*		
Main Emphasis	*Music Recording*	**Program is**	*Technical*
Accreditations	*Dept of Ed*		
Number of Studios	*3*	**Is school non-profit?**	*No*
Types of Studios	○ *Acoustic Research* ○ *Electronic Music* ○ *Film/Foley* ○ *Radio* ○ *Audio Research* ◉ *Music Recording* ○ *Television* ○ *Video*		
Types of Recording	◉ *Analog Multitrack* ◉ *MIDI Sequencing* ○ *Video* ○ *Digital Multitrack* ○ *DAW* ○ *Film*		
Other Resources	○ *Professional Studios* ○ *Television Stations* ○ *Radio Stations* ○ *Theater Tech Dept*		
Class Size Lecture	*12-20*	**Class Size Lab**	*1*
Assistance	○ *Housing* ◉ *Scholarships* ◉ *Internships* ○ *Financial Aid* ○ *Work-study* ◉ *Job Placement*		
Admission Policy	*Open*	**Language**	*English and Dutch*
Prerequisites	○ *High School Diploma* ○ *SAT-ACT* ○ *Music Audition* *16 years of age, hearing & pretest*		
Industry Affiliations	◉ *AES* ○ *ASA* ○ *NACB* ○ *NARAS* ○ *SMPTE* ○ *NAMBI* ○ *APRS* ○ *NAB* ○ *NAMM* ○ *SBE* ○ *SPARS* ○ *MEIEA*		
Fulltime Faculty	*3*	**Parttime Faculty**	*4*
Faculty Awards			
Program Awards			
Student Awards			
Research areas	*See School of Audio Engineering- Adelaide for profile*		

School of Audio Engineering- Berlin

Address	*Seestr. 64* *Berlin 65* *1000*	*Germany*	
Phone	*030-456-5137*	**FAX**	*030-456-5173*
Director	*Tom Misner*	**Admission Contact**	*Thomas Nommensen*
Program Founded	*1976*	**School Type**	*Trade School*
Program Offered	*Audio Engineer, Tonmeister, Studio Assistant*		
Degrees Offered	*Diploma, Degree, Certificate*		
Program Length	*18 months, 12 months, 6 months*		
Estimated Tuition	*$300 per month*		
Main Emphasis	*Music Recording*	**Program is**	*Technical*
Accreditations	*Dept of Ed*		
Number of Studios	*3*	**Is school non-profit?**	*No*
Types of Studios	○ *Acoustic Research* ○ *Electronic Music* ○ *Film/Foley* ○ *Radio* ○ *Audio Research* ◉ *Music Recording* ○ *Television* ○ *Video*		
Types of Recording	◉ *Analog Multitrack* ◉ *MIDI Sequencing* ○ *Video* ○ *Digital Multitrack* ○ *DAW* ○ *Film*		
Other Resources	○ *Professional Studios* ○ *Television Stations* ○ *Radio Stations* ○ *Theater Tech Dept*		
Class Size Lecture	*12-22*	**Class Size Lab**	*1*
Assistance	○ *Housing* ◉ *Scholarships* ◉ *Internships* ○ *Financial Aid* ○ *Work-study* ◉ *Job Placement*		
Admission Policy	*Open*	**Language**	*German*
Prerequisites	○ *High School Diploma* ○ *SAT-ACT* ○ *Music Audition* *16 years of age, hearing & pretest*		
Industry Affiliations	◉ *AES* ○ *ASA* ○ *NACB* ○ *NARAS* ○ *SMPTE* ○ *NAMBI* ○ *APRS* ○ *NAB* ○ *NAMM* ○ *SBE* ○ *SPARS* ○ *MEIEA*		
Fulltime Faculty	*5*	**Parttime Faculty**	*4*
Faculty Awards			
Program Awards			
Student Awards			
Research areas	*See School of Audio Engineering- Adelaide for profile*		

School of Audio Engineering- Brisbane

Address	*22 Heussler Tce, Milton* *Brisbane* *4064*	*Queensland* *Australia*	
Phone	*61-7-369-8108*	**FAX**	*61-2-211-3308*
Director	*Tom Misner*	**Admission Contact**	*Michael Quinn*
Program Founded	*1976*	**School Type**	*Trade School*
Program Offered	*Audio Engineer, Studio Assistant, Live Sound*		
Degrees Offered	*Diploma, Certificate, Certificate*		
Program Length	*15 months, 6 months, 6 months*		
Estimated Tuition	*$300 per month*		
Main Emphasis	*Music Recording*	**Program is**	*Technical*
Accreditations	*Dept of Ed*		
Number of Studios	*1*	**Is school non-profit?**	*No*
Types of Studios	○ *Acoustic Research* ○ *Electronic Music* ○ *Film/Foley* ○ *Radio* ○ *Audio Research* ◉ *Music Recording* ○ *Television* ○ *Video*		
Types of Recording	◉ *Analog Multitrack* ◉ *MIDI Sequencing* ○ *Video* ○ *Digital Multitrack* ○ *DAW* ○ *Film*		
Other Resources	○ *Professional Studios* ○ *Television Stations* ○ *Radio Stations* ○ *Theater Tech Dept*		
Class Size Lecture	*20-22*	**Class Size Lab**	*1*
Assistance	○ *Housing* ◉ *Scholarships* ◉ *Internships* ○ *Financial Aid* ○ *Work-study* ◉ *Job Placement*		
Admission Policy	*Open*	**Language**	*English*
Prerequisites	○ *High School Diploma* ○ *SAT-ACT* ○ *Music Audition* *16 years of age, hearing & pretest*		
Industry Affiliations	◉ *AES* ○ *ASA* ○ *NACB* ○ *NARAS* ○ *SMPTE* ○ *NAMBI* ○ *APRS* ○ *NAB* ○ *NAMM* ○ *SBE* ○ *SPARS* ○ *MEIEA*		
Fulltime Faculty	*2*	**Parttime Faculty**	*3*
Faculty Awards			
Program Awards			
Student Awards			
Research areas	*See School of Audio Engineering- Adelaide for profile*		

School of Audio Engineering- Frankfurt

Address *Homburger Landstrasse 182*
Frankfurt 50
6000 *Germany*

Phone *069-54-32-62* **FAX** *069-548 44 43*

Director *Tom Misner* **Admission Contact** *Bernhard*

Program Founded *1976* **School Type** *Trade School*

Program Offered *Audio Engineer, Tonmeister, Studio Assistant*

Degrees Offered *Diploma, Degree, Certificate*

Program Length *18 months, 12 months, 6 months*

Estimated Tuition *$300 per month*

Main Emphasis *Music Recording* **Program is** *Technical*

Accreditations *Dept of Ed*

Number of Studios *3* **Is school non-profit?** *No*

Types of Studios ○ *Acoustic Research* ○ *Electronic Music* ○ *Film/Foley* ○ *Radio*
○ *Audio Research* ◉ *Music Recording* ○ *Television* ○ *Video*

Types of Recording ◉ *Analog Multitrack* ◉ *MIDI Sequencing* ○ *Video*
○ *Digital Multitrack* ○ *DAW* ○ *Film*

Other Resources ○ *Professional Studios* ○ *Television Stations*
○ *Radio Stations* ○ *Theater Tech Dept*

Class Size Lecture *12-24* **Class Size Lab** *1*

Assistance ○ *Housing* ◉ *Scholarships* ◉ *Internships*
○ *Financial Aid* ○ *Work-study* ◉ *Job Placement*

Admission Policy *Open* **Language** *German*

Prerequisites ○ *High School Diploma* ○ *SAT-ACT* ○ *Music Audition*
16 years of age, hearing & pretest

Industry Affiliations ◉ *AES* ○ *ASA* ○ *NACB* ○ *NARAS* ○ *SMPTE* ○ *NAMBI*
○ *APRS* ○ *NAB* ○ *NAMM* ○ *SBE* ○ *SPARS* ○ *MEIEA*

Fulltime Faculty *5* **Parttime Faculty** *4*

Faculty Awards

Program Awards

Student Awards

Research areas *See School of Audio Engineering- Adelaide for profile*

School of Audio Engineering- Glasgow

Address *33 Coatbank Street*
Coatbridge *Glasgow*
ML5 3SP *Scotland*

Phone *0236-436561* **FAX**

Director *Tom Misner* **Admission Contact** *Gordon McMillan*

Program Founded *1976* **School Type** *Trade School*

Program Offered *Audio Engineer, Tonmeister, Studio Assistant, Production*

Degrees Offered *Diploma, Degree, Certificate, Certificate*

Program Length *15 months, 12 months, 6 months, 3 months*

Estimated Tuition *$300 per month*

Main Emphasis *Music Recording* **Program is** *Technical*

Accreditations *Dept of Ed*

Number of Studios *3* **Is school non-profit?** *No*

Types of Studios ○ *Acoustic Research* ○ *Electronic Music* ○ *Film/Foley* ○ *Radio*
○ *Audio Research* ● *Music Recording* ○ *Television* ○ *Video*

Types of Recording ● *Analog Multitrack* ● *MIDI Sequencing* ○ *Video*
○ *Digital Multitrack* ○ *DAW* ○ *Film*

Other Resources ● *Professional Studios* ○ *Television Stations*
○ *Radio Stations* ○ *Theater Tech Dept*

Class Size Lecture *12-22* **Class Size Lab** *1-2*

Assistance ○ *Housing* ● *Scholarships* ● *Internships*
○ *Financial Aid* ○ *Work-study* ● *Job Placement*

Admission Policy *Open* **Language** *English*

Prerequisites ○ *High School Diploma* ○ *SAT-ACT* ○ *Music Audition*
16 years of age, hearing & pretest

Industry Affiliations ● *AES* ○ *ASA* ○ *NACB* ○ *NARAS* ○ *SMPTE* ○ *NAMBI*
○ *APRS* ○ *NAB* ○ *NAMM* ○ *SBE* ○ *SPARS* ○ *MEIEA*

Fulltime Faculty *7* **Parttime Faculty** *9*

Faculty Awards

Program Awards

Student Awards

Research areas *See School of Audio Engineering- Adelaide for profile*

School of Audio Engineering- Kuala Lumpur

Address	*Lot 5, Jalan 13/2 Petaling Jaya* *Kuala Lumpur* *Selangor* *46200* *Malaysia*
Phone	*03-756-7212* **FAX** *03-757-2650*
Director	*Tom Misner* **Admission Contact** *Darien Nagle*
Program Founded	*1976* **School Type** *Trade School*
Program Offered	*Audio Engineer, Tonmeister, Studio Assistant*
Degrees Offered	*Diploma, Degree, Certificate*
Program Length	*15 months, 12 months, 6 months*
Estimated Tuition	*$300 per month*
Main Emphasis	*Music Recording* **Program is** *Technical*
Accreditations	*Dept of Ed*
Number of Studios	*3* **Is school non-profit?** *No*
Types of Studios	O *Acoustic Research* O *Electronic Music* O *Film/Foley* O *Radio* O *Audio Research* ◉ *Music Recording* O *Television* O *Video*
Types of Recording	◉ *Analog Multitrack* ◉ *MIDI Sequencing* O *Video* O *Digital Multitrack* O *DAW* O *Film*
Other Resources	◉ *Professional Studios* O *Television Stations* O *Radio Stations* O *Theater Tech Dept*
Class Size Lecture	*12-24* **Class Size Lab** *1*
Assistance	O *Housing* ◉ *Scholarships* ◉ *Internships* O *Financial Aid* O *Work-study* ◉ *Job Placement*
Admission Policy	*Open* **Language** *English and Malay*
Prerequisites	O *High School Diploma* O *SAT-ACT* O *Music Audition* *16 years of age, hearing & pretest*
Industry Affiliations	◉ *AES* O *ASA* O *NACB* O *NARAS* O *SMPTE* O *NAMBI* O *APRS* O *NAB* O *NAMM* O *SBE* O *SPARS* O *MEIEA*
Fulltime Faculty	*4* **Parttime Faculty** *3*
Faculty Awards	
Program Awards	
Student Awards	
Research areas	*See School of Audio Engineering- Adelaide for profile*

School of Audio Engineering- London

Address *United House, North Road*
London *England*
N7 9DP *UK*

Phone *44-71-6092653* **FAX** *44-71-6096944*

Director *Tom Minser* **Admission Contact** *Michael Pollard*

Program Founded *1976* **School Type** *Trade School*

Program Offered *Audio Engineer, Tonmeister, Studio Assistant, Production*

Degrees Offered *Diploma, Degree, Certificate, Certificate*

Program Length *15 months, 12 months, 6 months, 3 months*

Estimated Tuition *$300 per month*

Main Emphasis *Music Recording* **Program is** *Technical*

Accreditations *Dept of Ed*

Number of Studios *3* **Is school non-profit?** *No*

Types of Studios
○ *Acoustic Research* ○ *Electronic Music* ○ *Film/Foley* ○ *Radio*
○ *Audio Research* ◉ *Music Recording* ○ *Television* ○ *Video*

Types of Recording
◉ *Analog Multitrack* ◉ *MIDI Sequencing* ○ *Video*
○ *Digital Multitrack* ○ *DAW* ○ *Film*

Other Resources
◉ *Professional Studios* ○ *Television Stations*
○ *Radio Stations* ○ *Theater Tech Dept*

Class Size Lecture *12-20* **Class Size Lab** *1-2*

Assistance
○ *Housing* ◉ *Scholarships* ◉ *Internships*
○ *Financial Aid* ○ *Work-study* ◉ *Job Placement*

Admission Policy *Open* **Language** *English*

Prerequisites ○ *High School Diploma* ○ *SAT-ACT* ○ *Music Audition*
16 years of age, hearing & pretest

Industry Affiliations
◉ *AES* ○ *ASA* ○ *NACB* ○ *NARAS* ○ *SMPTE* ○ *NAMBI*
○ *APRS* ○ *NAB* ○ *NAMM* ○ *SBE* ○ *SPARS* ○ *MEIEA*

Fulltime Faculty *7* **Parttime Faculty** *9*

Faculty Awards

Program Awards

Student Awards

Research areas *See School of Audio Engineering- Adelaide for profile*

School of Audio Engineering- Melbourne

Address	*80-86 Inkerman Street, St. Kilda* *Melbourne* *Victoria* *3182* *Australia*
Phone	*61-3-534-4403* **FAX** *61-3-525-3542*
Director	*Tom Misner* **Admission Contact** *Tony Corr*
Program Founded	*1976* **School Type** *Trade School*
Program Offered	*Audio Engineer, Studio Assistant, Live Sound*
Degrees Offered	*Diploma, Certificate, Certificate*
Program Length	*15 months, 6 months, 6 months*
Estimated Tuition	*$300 per month*
Main Emphasis	*Music Recording* **Program is** *Technical*
Accreditations	*Dept of Ed*
Number of Studios	*2* **Is school non-profit?** *No*
Types of Studios	○ *Acoustic Research* ○ *Electronic Music* ○ *Film/Foley* ○ *Radio* ○ *Audio Research* ○ *Music Recording* ○ *Television* ○ *Video*
Types of Recording	◉ *Analog Multitrack* ○ *MIDI Sequencing* ○ *Video* ○ *Digital Multitrack* ○ *DAW* ○ *Film*
Other Resources	○ *Professional Studios* ○ *Television Stations* ○ *Radio Stations* ○ *Theater Tech Dept*
Class Size Lecture	*20-22* **Class Size Lab** *1*
Assistance	○ *Housing* ◉ *Scholarships* ◉ *Internships* ○ *Financial Aid* ○ *Work-study* ◉ *Job Placement*
Admission Policy	*Open* **Language** *English*
Prerequisites	○ *High School Diploma* ○ *SAT-ACT* ○ *Music Audition* *16 years of age, hearing & pretest*
Industry Affiliations	◉ *AES* ○ *ASA* ○ *NACB* ○ *NARAS* ○ *SMPTE* ○ *NAMBI* ○ *APRS* ○ *NAB* ○ *NAMM* ○ *SBE* ○ *SPARS* ○ *MEIEA*
Fulltime Faculty	*3* **Parttime Faculty** *2*
Faculty Awards	
Program Awards	
Student Awards	
Research areas	*See School of Audio Engineering- Adelaide for profile*

School of Audio Engineering- Munich

Address	*Hofer Str. 3 Munich 83 8000*	*Germany*	
Phone	*089-67-51-67*	**FAX**	*089-6701811*
Director	*Tom Misner*	**Admission Contact**	*Rudy Grieme*
Program Founded	*1977*	**School Type**	*Trade School*
Program Offered	*Audio Engineer, Tonmeister, Studio Assistant*		
Degrees Offered	*Diploma, Degree, Certificate*		
Program Length	*18 months, 12 months, 6 months*		
Estimated Tuition	*$300 per month*		
Main Emphasis	*Music Recording*	**Program is**	*Technical*
Accreditations	*Dept of Ed*		
Number of Studios	*3*	**Is school non-profit?**	*No*
Types of Studios	○ *Acoustic Research* ○ *Electronic Music* ○ *Film/Foley* ○ *Radio* ○ *Audio Research* ◉ *Music Recording* ○ *Television* ○ *Video*		
Types of Recording	◉ *Analog Multitrack* ◉ *MIDI Sequencing* ○ *Video* ○ *Digital Multitrack* ○ *DAW* ○ *Film*		
Other Resources	◉ *Professional Studios* ○ *Television Stations* ○ *Radio Stations* ○ *Theater Tech Dept*		
Class Size Lecture	*12-24*	**Class Size Lab**	*1*
Assistance	○ *Housing* ◉ *Scholarships* ◉ *Internships* ○ *Financial Aid* ○ *Work-study* ◉ *Job Placement*		
Admission Policy	*Open*	**Language**	*German*
Prerequisites	○ *High School Diploma* ○ *SAT-ACT* ○ *Music Audition* *16 years of age, hearing & pretest*		
Industry Affiliations	◉ *AES* ○ *ASA* ○ *NACB* ○ *NARAS* ○ *SMPTE* ○ *NAMBI* ○ *APRS* ○ *NAB* ○ *NAMM* ○ *SBE* ○ *SPARS* ○ *MEIEA*		
Fulltime Faculty	*5*	**Parttime Faculty**	*4*
Faculty Awards			
Program Awards			
Student Awards			
Research areas	*See School of Audio Engineering- Adelaide for profile*		

School of Audio Engineering- Paris

Address *33, Rue le la Porte d'Aubervillers*
Paris
75019 *France*

Phone *43-1-330-4133* **FAX** *43-1-330-4135*

Director *Tom Misner* **Admission Contact** *Michael Brück*

Program Founded *1976* **School Type** *Trade School*

Program Offered *Audio Engineer, Tonmeister, Studio Assistant*

Degrees Offered *Diploma, Degree, Certificate*

Program Length *15 months, 12 months, 6 months*

Estimated Tuition *$300 per month*

Main Emphasis *Music Recording* **Program is** *Technical*

Accreditations *Dept of Ed*

Number of Studios *3* **Is school non-profit?** *No*

Types of Studios
- ○ *Acoustic Research*
- ○ *Electronic Music*
- ○ *Film/Foley*
- ○ *Radio*
- ○ *Audio Research*
- ◉ *Music Recording*
- ○ *Television*
- ○ *Video*

Types of Recording
- ◉ *Analog Multitrack*
- ◉ *MIDI Sequencing*
- ○ *Video*
- ○ *Digital Multitrack*
- ○ *DAW*
- ○ *Film*

Other Resources
- ◉ *Professional Studios*
- ○ *Television Stations*
- ○ *Radio Stations*
- ○ *Theater Tech Dept*

Class Size Lecture *22* **Class Size Lab** *1*

Assistance
- ○ *Housing*
- ○ *Scholarships*
- ◉ *Internships*
- ◉ *Financial Aid*
- ○ *Work-study*
- ◉ *Job Placement*

Admission Policy *Open* **Language** *French*

Prerequisites
- ○ *High School Diploma*
- ○ *SAT-ACT*
- ○ *Music Audition*

16 years of age, hearing & pretest

Industry Affiliations
- ◉ *AES*
- ○ *ASA*
- ○ *NACB*
- ○ *NARAS*
- ○ *SMPTE*
- ○ *NAMBI*
- ○ *APRS*
- ○ *NAB*
- ○ *NAMM*
- ○ *SBE*
- ○ *SPARS*
- ○ *MEIEA*

Fulltime Faculty *6* **Parttime Faculty** *6*

Faculty Awards

Program Awards

Student Awards

Research areas *See School of Audio Engineering- Adelaide for profile*

School of Audio Engineering- Perth

Address *42 Wickham Street*
East Perth *Western Australia*
6000 *Australia*

Phone *09-325-4533* **FAX** *09-325-4533*

Director *Tom Misner* **Admission Contact** *Dale Blond*

Program Founded *1976* **School Type** *Trade School*

Program Offered *Audio Engineer, Studio Assistant, Live Sound*

Degrees Offered *Diploma, Certificate, Certificate*

Program Length *15 months, 6 months, 6 months*

Estimated Tuition *$300 per month*

Main Emphasis *Music Recording* **Program is** *Technical*

Accreditations *Dept of Ed*

Number of Studios *2* **Is school non-profit?** *No*

Types of Studios O *Acoustic Research* O *Electronic Music* O *Film/Foley* O *Radio*
O *Audio Research* ◉ *Music Recording* O *Television* O *Video*

Types of Recording ◉ *Analog Multitrack* ◉ *MIDI Sequencing* O *Video*
O *Digital Multitrack* O *DAW* O *Film*

Other Resources ◉ *Professional Studios* O *Television Stations*
O *Radio Stations* O *Theater Tech Dept*

Class Size Lecture *20-23* **Class Size Lab** *1*

Assistance O *Housing* ◉ *Scholarships* O *Internships*
O *Financial Aid* O *Work-study* O *Job Placement*

Admission Policy *Open* **Language** *English*

Prerequisites O *High School Diploma* O *SAT-ACT* O *Music Audition*
16 years of age, hearing & pretest

Industry Affiliations ◉ *AES* O *ASA* O *NACB* O *NARAS* O *SMPTE* O *NAMBI*
O *APRS* O *NAB* O *NAMM* O *SBE* O *SPARS* O *MEIEA*

Fulltime Faculty *2* **Parttime Faculty** *3*

Faculty Awards

Program Awards

Student Awards

Research areas *See School of Audio Engineering- Adelaide for profile*

School of Audio Engineering- Singapore

Address	*122 Middle Road, #04-08 Midlink Plaza* *Singapore* *0716* *Singapore*
Phone	*65-3652523* **FAX** *65-3652524*
Director	*Tom Misner* **Admission Contact** *George Moik*
Program Founded	*1976* **School Type** *Trade School*
Program Offered	*Audio Engineer, Tonmeister, Studio Assistant*
Degrees Offered	*Diploma, Degree, Certificate*
Program Length	*15 months, 12 months, 6 months*
Estimated Tuition	*$300 per month*
Main Emphasis	*Music Recording* **Program is** *Technical*
Accreditations	*Dept of Ed*
Number of Studios	*2* **Is school non-profit?** *No*
Types of Studios	○ *Acoustic Research* ○ *Electronic Music* ○ *Film/Foley* ○ *Radio* ○ *Audio Research* ◉ *Music Recording* ○ *Television* ○ *Video*
Types of Recording	◉ *Analog Multitrack* ◉ *MIDI Sequencing* ○ *Video* ○ *Digital Multitrack* ○ *DAW* ○ *Film*
Other Resources	○ *Professional Studios* ○ *Television Stations* ○ *Radio Stations* ○ *Theater Tech Dept*
Class Size Lecture	*12-24* **Class Size Lab** *1*
Assistance	○ *Housing* ○ *Scholarships* ○ *Internships* ○ *Financial Aid* ○ *Work-study* ○ *Job Placement*
Admission Policy	*Open* **Language** *English*
Prerequisites	○ *High School Diploma* ○ *SAT-ACT* ○ *Music Audition* *16 years of age, hearing & pretest*
Industry Affiliations	◉ *AES* ○ *ASA* ○ *NACB* ○ *NARAS* ○ *SMPTE* ○ *NAMBI* ○ *APRS* ○ *NAB* ○ *NAMM* ○ *SBE* ○ *SPARS* ○ *MEIEA*
Fulltime Faculty	*3* **Parttime Faculty** *2*
Faculty Awards	
Program Awards	
Student Awards	
Research areas	*See School of Audio Engineering- Adelaide for profile*

School of Audio Engineering- Sydney

Address	*68-72 Wentworth Avenue* *Sydney* *2010*	*New South Wales* *Australia*	
Phone	*61-2-21-13711*	**FAX**	*61-2-211-3308*
Director	*Tom Misner*	**Admission Contact**	*Angela Marino*
Program Founded	*1976*	**School Type**	*Trade School*
Program Offered	*Audio Engineer, Tonmeister, Studio Assistant, Video Production*		
Degrees Offered	*Diploma, Degree, Certificate, Degree*		
Program Length	*15 months, 12 months, 6 months, 12 months*		
Estimated Tuition	*$300 per month*		
Main Emphasis	*Music Recording*	**Program is**	*Technical*
Accreditations	*Dept of Ed*		
Number of Studios	*3*	**Is school non-profit?**	*No*
Types of Studios	○ *Acoustic Research* ○ *Electronic Music* ○ *Film/Foley* ○ *Radio* ○ *Audio Research* ◉ *Music Recording* ○ *Television* ○ *Video*		
Types of Recording	◉ *Analog Multitrack* ◉ *MIDI Sequencing* ○ *Video* ○ *Digital Multitrack* ○ *DAW* ○ *Film*		
Other Resources	○ *Professional Studios* ○ *Television Stations* ○ *Radio Stations* ○ *Theater Tech Dept*		
Class Size Lecture	*9-22*	**Class Size Lab**	*1-2*
Assistance	○ *Housing* ◉ *Scholarships* ○ *Internships* ○ *Financial Aid* ○ *Work-study* ◉ *Job Placement*		
Admission Policy	*Open*	**Language**	*English*
Prerequisites	○ *High School Diploma* ○ *SAT-ACT* ○ *Music Audition* *16 years of age, hearing & pretest*		
Industry Affiliations	◉ *AES* ○ *ASA* ○ *NACB* ○ *NARAS* ○ *SMPTE* ○ *NAMBI* ○ *APRS* ○ *NAB* ○ *NAMM* ○ *SBE* ○ *SPARS* ○ *MEIEA*		
Fulltime Faculty	*9*	**Parttime Faculty**	*11*
Faculty Awards			
Program Awards			
Student Awards			
Research areas	*See School of Audio Engineering- Adelaide for profile*		

School of Audio Engineering- Vienna

Address *Leystr. 43*
Vienna
A-1200 *Austria*

Phone *0222-3304133* **FAX** *0222-3304135*

Director *Tom Misner* **Admission Contact** *Michael Brück*

Program Founded *1976* **School Type** *Trade School*

Program Offered *Audio Engineer, Tonmeister, Studio Assistant*

Degrees Offered *Diploma, Degree, Certificate*

Program Length *18 months, 12 months, 6 months*

Estimated Tuition *$300 per month*

Main Emphasis *Music Recording* **Program is** *Technical*

Accreditations *Dept of Ed*

Number of Studios *3* **Is school non-profit?** *No*

Types of Studios
O *Acoustic Research* O *Electronic Music* O *Film/Foley* O *Radio*
O *Audio Research* ◉ *Music Recording* O *Television* O *Video*

Types of Recording
◉ *Analog Multitrack* ◉ *MIDI Sequencing* O *Video*
O *Digital Multitrack* O *DAW* O *Film*

Other Resources
◉ *Professional Studios* O *Television Stations*
O *Radio Stations* O *Theater Tech Dept*

Class Size Lecture *12-24* **Class Size Lab** *1*

Assistance
O *Housing* ◉ *Scholarships* ◉ *Internships*
O *Financial Aid* O *Work-study* ◉ *Job Placement*

Admission Policy *Open* **Language** *German*

Prerequisites
O *High School Diploma* O *SAT-ACT* O *Music Audition*
16 years of age, hearing & pretest

Industry Affiliations
◉ *AES* O *ASA* O *NACB* O *NARAS* O *SMPTE* O *NAMBI*
O *APRS* O *NAB* O *NAMM* O *SBE* O *SPARS* O *MEIEA*

Fulltime Faculty *5* **Parttime Faculty** *4*

Faculty Awards

Program Awards

Student Awards

Research areas *See School of Audio Engineering- Adelaide for profile*

South Plains College

Address *1401 S College Avenue*
Levelland *TX*
79336 *USA*

Phone *806-894-9611* **FAX** *806-894-5274*

Director *J Stoddard* **Admission Contact** *Bobby James*

Program Founded *1980* **School Type** *Community College*

Program Offered *Sound Technology, Performing Arts Production Technology*

Degrees Offered *Associate in Applied Science*

Program Length *2 years*

Estimated Tuition *$3,000*

Main Emphasis *Audio Engineering* **Program is** *Technical*

Accreditations *State of Texas*

Number of Studios *5* **Is school non-profit?** *Yes*

Types of Studios
○ *Acoustic Research* ◉ *Electronic Music* ○ *Film/Foley* ○ *Radio*
○ *Audio Research* ◉ *Music Recording* ○ *Television* ◉ *Video*

Types of Recording
◉ *Analog Multitrack* ◉ *MIDI Sequencing* ◉ *Video*
◉ *Digital Multitrack* ○ *DAW* ○ *Film*

Other Resources
◉ *Professional Studios* ◉ *Television Stations*
○ *Radio Stations* ○ *Theater Tech Dept*

Class Size Lecture *12* **Class Size Lab** *12*

Assistance
◉ *Housing* ◉ *Scholarships* ○ *Internships*
◉ *Financial Aid* ◉ *Work-study* ○ *Job Placement*

Admission Policy *Open* **Language** *English*

Prerequisites ◉ *High School Diploma* ◉ *SAT-ACT* ○ *Music Audition*

Industry Affiliations
◉ *AES* ○ *ASA* ○ *NACB* ◉ *NARAS* ○ *SMPTE* ○ *NAMBI*
○ *APRS* ◉ *NAB* ○ *NAMM* ○ *SBE* ○ *SPARS* ○ *MEIEA*

Fulltime Faculty *5* **Parttime Faculty** *0*

Faculty Awards

Program Awards

Student Awards

Research areas

South Plains College

South Plains College offers two degree programs: Sound Technology and Performing Arts Production Technology. Both are two-year programs resulting Associate in Applied Science degrees. The Sound Technology program is designed to provide preparation and practical experience for students interested in entry-level positions as recording engineers/producers, sound reinforcement technicians, and broadcast audio specialists. The curriculum includes courses in electronics, equipment operations and maintenance, music theory, acoustics, business, and more.

The Performing Arts Production Technology program is designed for students interested in entry-level technical production and event management positions in video production facilities, entertainment venues, and theaters. The program provides training in audio and video production, television, film, stage and concert lighting, stagecrafts, and business management.

Both programs are housed in the Tom T. Hall Recording and Production Studio, a multi-media production center featuring a sound stage, complete video control suite, a professional lighting system, and an automated multitrack recording facility. Students receive hands-on experience through music recording projects and production of music videos, commercials, musical shows, dinner theaters, and fashion shows.

Southern Ohio College

Address	*1055 Laidlaw Avenue* *Cincinnati* *45237*	*OH* *USA*	
Phone	*512-242-3791*	**FAX**	*513-242-2844*
Director	*Mark Turner*	**Admission Contact**	*Lauren Smith*
Program Founded	*1978*	**School Type**	*Community College*
Program Offered	*Audio Video Production*		
Degrees Offered	*Associate of Applied Business*		
Program Length	*2 years*		
Estimated Tuition	*$1,380 per quarter*		
Main Emphasis	*Audio Engineering*	**Program is**	*Technical*
Accreditations	*NCASC, CCA, AICS*		
Number of Studios	*7*	**Is school non-profit?**	*No*

Types of Studios
- ○ *Acoustic Research*
- ○ *Electronic Music*
- ◉ *Film/Foley*
- ○ *Radio*
- ○ *Audio Research*
- ◉ *Music Recording*
- ◉ *Television*
- ◉ *Video*

Types of Recording
- ◉ *Analog Multitrack*
- ◉ *MIDI Sequencing*
- ◉ *Video*
- ○ *Digital Multitrack*
- ○ *DAW*
- ○ *Film*

Other Resources
- ○ *Professional Studios*
- ○ *Television Stations*
- ○ *Radio Stations*
- ○ *Theater Tech Dept*

Class Size Lecture *20* **Class Size Lab** *15*

Assistance
- ○ *Housing*
- ◉ *Scholarships*
- ◉ *Internships*
- ◉ *Financial Aid*
- ◉ *Work-study*
- ◉ *Job Placement*

Admission Policy *Selective* **Language** *English*

Prerequisites
- ◉ *High School Diploma*
- ○ *SAT-ACT*
- ○ *Music Audition*

Industry Affiliations
- ◉ *AES*
- ○ *ASA*
- ○ *NACB*
- ○ *NARAS*
- ○ *SMPTE*
- ○ *NAMBI*
- ○ *APRS*
- ○ *NAB*
- ○ *NAMM*
- ○ *SBE*
- ○ *SPARS*
- ○ *MEIEA*

Fulltime Faculty *2* **Parttime Faculty** *3*

Faculty Awards

Program Awards

Student Awards *Ohio Blue Chips*

Research areas

Southern Ohio College

Southern Ohio College's associate degree program in Audio Video Production provides graduates with the skills necessary to enter the fields of corporate and industrial audio and video communications as well as other non-broadcast production environments. The curriculum encompasses general education, industry-related business and computer courses, and specialized courses in all phases of audio and video production.

The Audio Video Production Department has an enrollment of 140 students and is housed on the main campus of Southern Ohio College in the heart of Cincinnati. Students are trained in all aspects of audio and video production in a hands-on environment. The department's state-of-the-industry facilities include two multitrack recording studios with ADR, Foley and SMPTE lock to video capabilities. Both studios offer an array of outboard gear, and the larger of the two has full MIDI capabilities and three isolation booths. There are an additional four studios for audio.

Video production is learned in both the department's multi-camera production studio and on-location utilizing three portable production units. The department has two computerized SMPTE on-line and two off-line video editing suites. Students also learn digital video effects, 2-D computer graphics, and multi-format video production.

Southern Ohio College maintains an open lab policy to allow students to utilize school facilities for personal projects. In fact, students are encourage to use the facilities whenever they are available.

Students hands-on experience is enhanced by real world projects. When they graduate they take with them a demo real which includes work for actual clients. Graduates from the Audio Video Production program are working in recording studio and production facilities in all aspects of the industry.

Southwest Texas State University

Address	*Music Department, 601 University Drive* *San Marcos* *TX* *78666* *USA*		
Phone	*512-245-2651*	**FAX**	
Director	*Mark Erickson*	**Admission Contact**	*Mark Erickson*
Program Founded	*1992*	**School Type**	*University*
Program Offered	*Sound Recording Technology*		
Degrees Offered	*Bachelor of Music*		
Program Length	*4 years*		
Estimated Tuition	*$6,488*		
Main Emphasis	*Audio Engineering*	**Program is**	*Technical*
Accreditations	*NASM*		
Number of Studios	*2*	**Is school non-profit?**	*Yes*
Types of Studios	○ *Acoustic Research* ◉ *Electronic Music* ○ *Film/Foley* ○ *Radio* ○ *Audio Research* ◉ *Music Recording* ○ *Television* ○ *Video*		
Types of Recording	◉ *Analog Multitrack* ◉ *MIDI Sequencing* ○ *Video* ◉ *Digital Multitrack* ◉ *DAW* ○ *Film*		
Other Resources	◉ *Professional Studios* ○ *Television Stations* ○ *Radio Stations* ○ *Theater Tech Dept*		
Class Size Lecture	*15*	**Class Size Lab**	*15*
Assistance	◉ *Housing* ◉ *Scholarships* ◉ *Internships* ◉ *Financial Aid* ◉ *Work-study* ◉ *Job Placement*		
Admission Policy	*Highly selective*	**Language**	*English*
Prerequisites	◉ *High School Diploma* ◉ *SAT-ACT* ◉ *Music Audition*		
Industry Affiliations	◉ *AES* ○ *ASA* ○ *NACB* ○ *NARAS* ○ *SMPTE* ○ *NAMBI* ○ *APRS* ○ *NAB* ○ *NAMM* ○ *SBE* ◉ *SPARS* ○ *MEIEA*		
Fulltime Faculty	*1*	**Parttime Faculty**	*2*
Faculty Awards			
Program Awards			
Student Awards			
Research areas			

Southwest Texas State University

Southwest Texas State University (SWT) is the only university in the state of Texas to offer a baccalaureate degree in Sound Recording Technology (SRT). Admission is competitive, with less than 15 freshman admitted annually. Applicants should have significant musical abilities, well-developed aural skills, and competencies in math and science. The curriculum emphasizes recording, music, math/science, general studies, and an internship. Graduates receive a Bachelor of Music degree with an emphasis in Sound Recording Technology.

SWT owns and operates the Fire Station, a multipurpose recording facility. Students participate in commercial recording sessions while pursuing their SRT degrees. Fire Station Studios is both a digital and analog multitrack recording studio and a television/film soundstage with an infinity cyclorama. The professional facilities are fully-equipped and spacious. The studios' credits include many famous commercial artists and musicians. Students also have access to the Music Department's Macintosh computer lab and electronic music lab, each containing MIDI devices, multitrack recorders, and personal computers.

Stanford University, Music Department

Address *Center for Computer Research in Music*
Stanford *CA*
94305-8180 *USA*

Phone *415-723-3811* **FAX** *415-723-8468*

Director *John Chowning* **Admission Contact** *John Planting*

Program Founded *1975* **School Type** *University*

Program Offered *Music, Science, & Technology, Computer-Based Music Theory & Acoustics*

Degrees Offered *Bachelor of Arts, Doctor of Music Arts, Doctor of Philosophy*

Program Length *4 years, graduate degrees vary*

Estimated Tuition *$13,000 per year*

Main Emphasis *Electronic Music* **Program is** *Technical*

Accreditations

Number of Studios *3* **Is school non-profit?** *Yes*

Types of Studios
○ *Acoustic Research* ◉ *Electronic Music* ○ *Film/Foley* ○ *Radio*
◉ *Audio Research* ◉ *Music Recording* ○ *Television* ○ *Video*

Types of Recording
○ *Analog Multitrack* ◉ *MIDI Sequencing* ○ *Video*
◉ *Digital Multitrack* ○ *DAW* ○ *Film*

Other Resources
○ *Professional Studios* ○ *Television Stations*
○ *Radio Stations* ○ *Theater Tech Dept*

Class Size Lecture *25* **Class Size Lab** *16*

Assistance
○ *Housing* ○ *Scholarships* ○ *Internships*
◉ *Financial Aid* ○ *Work-study* ○ *Job Placement*

Admission Policy *Highly selective* **Language** *English*

Prerequisites ○ *High School Diploma* ◉ *SAT-ACT* ○ *Music Audition*

Industry Affiliations
○ *AES* ○ *ASA* ○ *NACB* ○ *NARAS* ○ *SMPTE* ○ *NAMBI*
○ *APRS* ○ *NAB* ○ *NAMM* ○ *SBE* ○ *SPARS* ○ *MEIEA*

Fulltime Faculty *3* **Parttime Faculty** *0*

Faculty Awards

Program Awards

Student Awards

Research areas *Psychoacoustics, Physical Modeling, Signal Processing, Languages*

Stanford University, Music Department

The Stanford Center for Computer Research in Music and Acoustics (CCRMA) is an interdisciplinary facility where composers and researchers work together using computer-based technology as a new musical and artistic medium, and as a research tool. Stanford offers both undergraduate and graduate programs centered around work at CCRMA.

Areas of ongoing interest at CCRMA include: applications hardware and software, synthesis techniques and algorithms, signal processing, digital recording and editing, psychoacoustics and musical acoustics, applied pattern recognition and artificial intelligence, computer-aided music manuscript, composition, and real-time applications with small systems.

The CCRMA community consists of administrative and technical staff, faculty, research associates, graduate research assistant, graduate and undergraduate students, visiting scholars and composers, and industrial associates. Major departments actively represented at CCRMA include Music, Electrical Engineering, Computer Science, and Psychology.

Center activities include academic courses, seminars, small interest group meetings, summer workshops, and other presentations. Concerts of computer music are presented regularly with an outdoor computer music festival in July. In-house technical reports and recordings are available, and public demonstrations of the ongoing work at CCRMA are held monthly during the academic year.

Prospective graduate students especially interested in the work at CCRMA should apply to the degree program most closely aligned with their specific field of interest, such as Music, Computer Science, Electrical Engineering, Psychology, etc. Graduate degree programs offered in music are the DMA in Composition and the PhD in Computer-based Music Theory and Acoustics. Acceptance in music theory or composition is largely based upon musical criteria, not computing knowledge. All courses at CCRMA are also open to undergraduate students. In addition, undergraduates can pursue independent studies.

CCRMA offers a series of four-week summer workshops which are opened to all those wishing to apply.

SUNY Fredonia

Address *Mason Hall, School of Music*
Fredonia *NY*
14063 *USA*

Phone *716-673-3221* **FAX** *716-673-3397*

Director *Dave Kerzner* **Admission Contact** *Dave Kerzner*

Program Founded *1976* **School Type** *University*

Program Offered *Sound Recording Technology*

Degrees Offered *Bachelor of Science*

Program Length *4 years*

Estimated Tuition *$8,600*

Main Emphasis *Music Recording* **Program is** *Semi-technical*

Accreditations *NASM*

Number of Studios *5* **Is school non-profit?** *Yes*

Types of Studios ○ *Acoustic Research* ◉ *Electronic Music* ○ *Film/Foley* ○ *Radio*
○ *Audio Research* ◉ *Music Recording* ○ *Television* ○ *Video*

Types of Recording ◉ *Analog Multitrack* ◉ *MIDI Sequencing* ○ *Video*
○ *Digital Multitrack* ○ *DAW* ○ *Film*

Other Resources ◉ *Professional Studios* ○ *Television Stations*
○ *Radio Stations* ◉ *Theater Tech Dept*

Class Size Lecture *12* **Class Size Lab** *12*

Assistance ◉ *Housing* ◉ *Scholarships* ◉ *Internships*
◉ *Financial Aid* ◉ *Work-study* ◉ *Job Placement*

Admission Policy *Highly selective* **Language** *English*

Prerequisites ◉ *High School Diploma* ◉ *SAT-ACT* ◉ *Music Audition*

Industry Affiliations ◉ *AES* ○ *ASA* ○ *NACB* ○ *NARAS* ○ *SMPTE* ○ *NAMBI*
○ *APRS* ○ *NAB* ○ *NAMM* ○ *SBE* ○ *SPARS* ○ *MEIEA*

Fulltime Faculty *3* **Parttime Faculty** *1*

Faculty Awards

Program Awards

Student Awards

Research areas

SUNY Fredonia

The Sound Recording Technology (SRT) program at SUNY Fredonia is an intensive, interdisciplinary program in audio engineering and related areas, including TV and radio, motion picture production, technical theater, and concert sound reinforcement. The program is a balance of liberal arts and sciences. The curriculum is based upon the central notion that an audio engineer is a musician with a broad range of scientific and engineering skills.

Students enrolling in the SRT program complete a rigorous four-year program that includes studio training, musical and scientific studies, and liberal arts and sciences coursework. SRT students are music majors and share the same basic educational responsibilities as all music majors.

Students can enroll either as freshman or transfer students. Entering freshman often have substantial musical experience, an aptitude for math and physics, and high grade point averages and SAT/ACT scores. Transfer students come from a variety of backgrounds- some already hold degrees and have substantial experience in music and performance. All have demonstrated ability and aptitude in the sciences, as well as very high motivation.

The three-year professional sequence of courses begins at the sophomore level and is designed to achieve a variety of goals: a working knowledge of the industry, training in recording production and live sound for different media, and experience in operations, maintenance, and management. In recognition of the volatile nature of the recording industry, courses are under constant revision. Emphasis is placed on developing the abilities of students to rapidly adapt to new conditions and new media systems. Students assist in the program's studios on a 24-hour basis and are expected to function in a professional manner.

An active student chapter of the Audio Engineering Society is on campus. The Sound Services Club provides sound reinforcement services for the campus community with a large participation of SRT students. Due to competition, applicants for the SRT program are encouraged to file paperwork and audition by March 1st for September admission.

Syracuse University

Address — Newhouse School: Radio, TV, Film Department
Syracuse NY
13244 USA

Phone 315-443-4004 **FAX** 315-443-3946

Director David Ruben **Admission Contact** Peter Moller

Program Founded 1974 **School Type** University

Program Offered Electronic Media Production

Degrees Offered Bachelor of Science, Master of Science

Program Length 4 years, 2 years

Estimated Tuition $60,000

Main Emphasis Broadcasting **Program is** Semi-technical

Accreditations Council for Ed in Journalism & Mass Comm

Number of Studios 10 **Is school non-profit?** Yes

Types of Studios
○ Acoustic Research ◉ Electronic Music ◉ Film/Foley ◉ Radio
○ Audio Research ◉ Music Recording ◉ Television ◉ Video

Types of Recording
◉ Analog Multitrack ◉ MIDI Sequencing ◉ Video
○ Digital Multitrack ◉ DAW ◉ Film

Other Resources
○ Professional Studios ○ Television Stations
◉ Radio Stations ◉ Theater Tech Dept

Class Size Lecture Varies **Class Size Lab** Varies

Assistance
◉ Housing ◉ Scholarships ◉ Internships
◉ Financial Aid ◉ Work-study ◉ Job Placement

Admission Policy Highly selective **Language** English

Prerequlsites ◉ High School Diploma ◉ SAT-ACT ○ Music Audition

Industry Affiliations
◉ AES ○ ASA ◉ NACB ○ NARAS ○ SMPTE ○ NAMBI
○ APRS ○ NAB ○ NAMM ○ SBE ○ SPARS ○ MEIEA

Fulltime Faculty Varies **Parttime Faculty** Varies

Faculty Awards

Program Awards

Student Awards

Research areas

Syracuse University

The Newhouse School of Communications at Syracuse University offers undergraduate and graduate degrees in Electronic Media Production. These programs prepare students for entry-level positions in electronic media as producers, designers, and directors of audio/video productions. Though no specific degree in audio is offered, students may assemble an emphasis in sound by taking advantage of flexible degree requirements and faculty advising. A variety of courses relating to audio are available from the School of Communications, the School of Music, the Art Media Studies Department, and the School of Engineering, to name a few.

The School of Communications and the Art Media Studies Department both offer courses in audio, computer imaging, video, and film. The School of Communications also offers classes related to commercial radio and television production, electronic news gathering, and broadcast journalism. Students benefit from a wide range of audio and video production facilities, as well as employment opportunities as three campus radio stations and a cable television station. Students can also gain sound reinforcement and concert staging experience with the University's Event Productions group.

The School of Music provides courses in music theory and performance, electronic music composition, and music business. Students can utilize a professionally equipped, multitrack recording studio, an extensive electronic music studio complete with analog and digital synthesis, a MIDI piano lab, and a computer-based ear training suite. A Music Industry Bachelor of Music or Minor in Music is available for those interested in music business. Plans are underway to create an audio track within the Music Industry program.

The Art Media Studies Department offers students creative opportunities for media art production. Based in the School of Art, the program includes computer graphics/art, film, and video art and emphasizes experimental applications of audio visual technology. A variety of studios support the program's activities, and student work is regularly featured at the University's professional art gallery.

Trebas Institute of Recording Arts

Address *6464 Sunset Boulevard*
Los Angeles *CA*
90028 *USA*

Phone *213-467-6800* **FAX**

Director *David P. Leonard* **Admission Contact** *David P. Leonard*

Program Founded *1979* **School Type** *Trade School*

Program Offered *Recording Arts & Sciences, Audio Engineering Technology, others*

Degrees Offered *Diploma*

Program Length *1 year*

Estimated Tuition *$9000*

Main Emphasis *Music Recording* **Program is** *Semi-technical*

Accreditations *ACCET*

Number of Studios *Varies* **Is school non-profit?** *No*

Types of Studios ○ *Acoustic Research* ◉ *Electronic Music* ○ *Film/Foley* ○ *Radio*
○ *Audio Research* ◉ *Music Recording* ○ *Television* ○ *Video*

Types of Recording ◉ *Analog Multitrack* ◉ *MIDI Sequencing* ○ *Video*
◉ *Digital Multitrack* ◉ *DAW* ○ *Film*

Other Resources ◉ *Professional Studios* ○ *Television Stations*
○ *Radio Stations* ○ *Theater Tech Dept*

Class Size Lecture *20-25* **Class Size Lab** *8*

Assistance ○ *Housing* ◉ *Scholarships* ◉ *Internships*
◉ *Financial Aid* ○ *Work-study* ◉ *Job Placement*

Admission Policy *Selective* **Language** *English*

Prerequisites ◉ *High School Diploma* ○ *SAT-ACT* ○ *Music Audition*
Interview, Entrance Exam

Industry Affiliations ◉ *AES* ○ *ASA* ○ *NACB* ○ *NARAS* ○ *SMPTE* ○ *NAMBI*
○ *APRS* ○ *NAB* ○ *NAMM* ○ *SBE* ◉ *SPARS* ◉ *MEIEA*

Fulltime Faculty *Varies* **Parttime Faculty** *Varies*

Faculty Awards

Program Awards *TEC Award nominess*

Student Awards

Research areas

Trebas Institute of Recording Arts

The Trebas Institute of Recording Arts operates four campuses in the United States and Canada which offer one-year programs in: Recording Arts and Sciences, Recorded Music Production, Audio Engineering Technology, and Music Business Administration. The Institute's goal is to provide comprehensive, college-level training and education in the music business and the recording arts. Their programs are designed to give students the knowledge, practical skills, and professionalism required to function successfully in the music, film, television, and video industries.

Trebas promotes comprehensive education in music technology, music business, and general communication skills. Students can choose from over 100 courses in 11 tracks of study, including audio engineering theory, computers, digital topics, electronics, music production, studio recording, general education, and others.

Trebas offers transfer credit between all of its campuses. Several scholarships, up to $5,000, are available at each campus. Government loans and grants are available for qualified students. Goldmark Memorial Scholarships are available for advanced studies. Limited internships are available, and job placement assistance is provided. International students are welcomed.

Facilities vary by campus, but most include: professional multitrack recording studios, electronic music/MIDI studios, electronics and computer labs, post-production mixing facilities, and mastering rooms. Students also benefit from the Institute's resource center and library, containing several thousand books and periodicals on music business, audio and video, recording, and music production.

Trebas graduates work with major record companies, recording studios, and performing artists.

The other Trebas Institute campuses are located in:

Montreal: Trebas Institute, 451 St Jean Street, Montreal, Quebec, H2Y 2R5, Canada, phone: 514-845-4141.

Toronto: Trebas Institute, 410 Dundas Street East, Toronto, Ontario, M5A 2A8, Canada, phone: 416-966-3066.

Vancouver: Trebas Institute, 112 East 3rd Avenue, Vancouver, British Columbia, V5T 1C8, Canada, phone: 604-872-2666.

Trod Nossel Recording, RIA

Address *10 George Street, PO Box 57*
Wallingford *CT*
06492 *USA*

Phone *203-269-4465* **FAX** *203-294-1745*

Director *Richard Robinson* **Admission Contact** *Tommy Cavalier*

Program Founded *1975* **School Type** *Workshop*

Program Offered *Modern Recording Techniques*

Degrees Offered *None*

Program Length *12 weeks*

Estimated Tuition *$450*

Main Emphasis *Music Recording* **Program is** *Semi-technical*

Accreditations

Number of Studios *2* **Is school non-profit?** *No*

Types of Studios ○ *Acoustic Research* ○ *Electronic Music* ○ *Film/Foley* ○ *Radio*
○ *Audio Research* ◉ *Music Recording* ○ *Television* ○ *Video*

Types of Recording ◉ *Analog Multitrack* ○ *MIDI Sequencing* ○ *Video*
◉ *Digital Multitrack* ○ *DAW* ○ *Film*

Other Resources ◉ *Professional Studios* ○ *Television Stations*
○ *Radio Stations* ○ *Theater Tech Dept*

Class Size Lecture *12* **Class Size Lab** *6*

Assistance ○ *Housing* ○ *Scholarships* ○ *Internships*
○ *Financial Aid* ○ *Work-study* ○ *Job Placement*

Admission Policy *Open* **Language** *English*

Prerequisites ○ *High School Diploma* ○ *SAT-ACT* ○ *Music Audition*

Industry Affiliations ◉ *AES* ○ *ASA* ○ *NACB* ○ *NARAS* ○ *SMPTE* ○ *NAMBI*
○ *APRS* ○ *NAB* ○ *NAMM* ○ *SBE* ◉ *SPARS* ○ *MEIEA*

Fulltime Faculty *3* **Parttime Faculty** *0*

Faculty Awards

Program Awards

Student Awards

Research areas

Trod Nossel Recording, RIA

The Recording Institute of America at Trod Nossel Studios offers a Modern Recording Techniques program. The course was started by musician/producer Vincent Testa who was the need for musicians to be educated in the basics of studio techniques without going through extensive electronics training.

The course is intended for musicians and home recordists who wish to gain an understanding of mutlitrack recording. Audiophiles, live sound engineers, and experienced recording engineers and producers in need of refreshing their audio recording fundamentals are welcomed.

The course consists of a minimum of 20 hours of theory and 20 hours of recording sessions spread over a 12-week period. There are demonstrations given in most theory sections, along with reading assignments and practical exercises to do at home. The practical, hands-on section of the course is a series of recording sessions with solo artists or bands.

An advanced course is available in which there are 75 hours of recording sessions over a 15-week period.

UCLA Extension

Address *10995 Le Conte Avenue, Room 437*
Los Angeles *CA*
90099-2137 *USA*

Phone *310-825-9064* **FAX** *310-206-7382*

Director *Van Webster* **Admission Contact** *Lisa Brewer Herring*

Program Founded *1977* **School Type** *University*

Program Offered *Recording Engineering, Recording Arts & Sciences, Electronic Music*

Degrees Offered *Certificate*

Program Length *2 years*

Estimated Tuition *Varies*

Main Emphasis *Electronic Music* **Program is** *Semi-technical*

Accreditations

Number of Studios *Varies* **Is school non-profit?** *Yes*

Types of Studios
○ *Acoustic Research* ◉ *Electronic Music* ◉ *Film/Foley* ◉ *Radio*
○ *Audio Research* ◉ *Music Recording* ◉ *Television* ◉ *Video*

Types of Recording
◉ *Analog Multitrack* ◉ *MIDI Sequencing* ◉ *Video*
○ *Digital Multitrack* ○ *DAW* ◉ *Film*

Other Resources
○ *Professional Studios* ○ *Television Stations*
○ *Radio Stations* ○ *Theater Tech Dept*

Class Size Lecture *20-40* **Class Size Lab** *6-10*

Assistance
○ *Housing* ◉ *Scholarships* ◉ *Internships*
◉ *Financial Aid* ○ *Work-study* ○ *Job Placement*

Admission Policy *Open* **Language** *English*

Prerequisites
○ *High School Diploma* ○ *SAT-ACT* ○ *Music Audition*
Physics, Electronics, Math Courses

Industry Affiliations
◉ *AES* ○ *ASA* ○ *NACB* ○ *NARAS* ○ *SMPTE* ○ *NAMBI*
○ *APRS* ○ *NAB* ○ *NAMM* ○ *SBE* ○ *SPARS* ○ *MEIEA*

Fulltime Faculty *Varies* **Parttime Faculty** *Varies*

Faculty Awards

Program Awards

Student Awards

Research areas

UCLA Extension

UCLA Extension offers a variety of continuing education programs through their Enterainment Studies program. Among those are certificate program in Recording Engineering, Recording Arts and Sciences, Electronic Music, Film Scoring, and Film, Television, and Video. Each curriculum is continually updated and expanded in consultation with industry experts, guilds and academies, professional associations, and university faculty. The certificate programs provide a structured approach to gaining perspective, knowledge, and training in the multi-faceted fields of the music industry.

The Certificate Program in Recording Engineering is a comprehensive training program offering a through study of music recording engineering taught in state-of-the-art facilities under the guidance of top industry professionals. The program encompasses both theory and practice in audio technology, equipment, musicianship, and business practices. The objective is to provide problem-solving techniques and skills that enable future engineers to meet the challenges of rapidly evolving technologies and the dynamic sound recording industry.

The Certificate in Recording Arts and Sciences is aimed at individuals seeking to advance their careers as artists, producers, engineers, managers, songwriters, and publishers. Founded in 1977, the program represents the first continuing education curriculum covering all aspects of the music industry.

The Certificate Program in Electronic Music allows students to study the use of electronic music technology for composition, performance, synthesis, and recording. The program benefits from the UCLA Extension Music Media Lab. The lab is equipped with student and instructor workstations, consisting of computer and synthesizer equipment from leading manufacturers. The lab is also available to students working in other programs.

UCLA Extension has the advantage of being in Los Angeles, major purveyor to the world of film, television, video, radio, and recorded music, and thereby a magnet for leading professionals in the artistic, business, legal, and technological disciplines of the industry. The school utilizes many of these professionals as faculty for lectures, seminars, and workshops.

A college or university degree is not required to enroll in any of the UCLA Extension courses, though some have certain prerequisites or require professional experience. Many college graduates use courses for continuing education or to complement their university studies. Students may enroll in individual courses without enrolling in a certificate program.

Unity Gain Recording Institute

Address	*2976-F Cleveland Avenue*		
	Fort Meyers	*FL*	
	33901	*USA*	
Phone	*813-332-4246*	**FAX**	
Director	*Anthony Iannucci*	**Admission Contact**	*Patricia Iannucci*
Program Founded	*1989*	**School Type**	*Trade School*
Program Offered	*Audio Recording*		
Degrees Offered	*Certificate*		
Program Length	*48 weeks, 144 hours*		
Estimated Tuition	*$1,730*		
Main Emphasis	*Audio Engineering*	**Program is**	*Technical*
Accreditations	*State of Florida*		
Number of Studios	*1*	**Is school non-profit?**	*No*
Types of Studios	○ *Acoustic Research* ○ *Electronic Music* ○ *Film/Foley* ○ *Radio* ○ *Audio Research* ◉ *Music Recording* ○ *Television* ○ *Video*		
Types of Recording	◉ *Analog Multitrack* ◉ *MIDI Sequencing* ○ *Video* ◉ *Digital Multitrack* ○ *DAW* ○ *Film*		
Other Resources	○ *Professional Studios* ○ *Television Stations* ○ *Radio Stations* ○ *Theater Tech Dept*		
Class Size Lecture	*5*	**Class Size Lab**	*5*
Assistance	○ *Housing* ○ *Scholarships* ◉ *Internships* ○ *Financial Aid* ○ *Work-study* ◉ *Job Placement*		
Admission Policy	*Selective*	**Language**	*English*
Prerequisites	◉ *High School Diploma* ○ *SAT-ACT* ○ *Music Audition*		
Industry Affiliations	◉ *AES* ○ *ASA* ○ *NACB* ○ *NARAS* ○ *SMPTE* ○ *NAMBI* ○ *APRS* ○ *NAB* ○ *NAMM* ○ *SBE* ○ *SPARS* ○ *MEIEA*		
Fulltime Faculty	*1*	**Parttime Faculty**	*1*
Faculty Awards			
Program Awards			
Student Awards			
Research areas			

Unity Gain Recording Institute

The Audio Recording program offered by Unity Gain is divided into four classes: Introduction to Audio Engineering, Advanced Techniques in Audio Engineering, MIDI Theory and Application, and Audio Recording Workshop. Each class consists of 36 hours of instruction spread over 12 weeks. Unity Gain also offers graduates additional studio time to gain additional experience and to build their resumes.

The Introduction to Audio Engineering class familiarizes students with sound basics, microphones, consoles, patchbays, and multitrack recorders. The Advanced Techniques class introduces equalization, special effects, noise reduction systems, and recorder alignment procedures. The MIDI class covers sequencers, synthesizers, samplers, computer languages, automation, notation, and virtual editing. The Audio Recording Workshop allows students to gain practical experience through weekly recording sessions.

University of California at Santa Cruz

Address *1156 High Street, Santa Cruz, CA 95064, USA*

Phone *408-459-2369* **FAX**

Director *Peter Elsea* **Admission Contact** *Director of Admissions*

Program Founded *1970* **School Type** *University*

Program Offered *Electronic Music Composition*

Degrees Offered *Bachelor of Arts, Master of Arts*

Program Length *4 years, 2 years*

Estimated Tuition *$14,000*

Main Emphasis *Electronic Music* **Program is** *Semi-technical*

Accreditations *WASC*

Number of Studios *3* **Is school non-profit?** *Yes*

Types of Studios
- ○ Acoustic Research
- ◉ Electronic Music
- ○ Film/Foley
- ○ Radio
- ○ Audio Research
- ○ Music Recording
- ○ Television
- ○ Video

Types of Recording
- ◉ Analog Multitrack
- ◉ MIDI Sequencing
- ○ Video
- ○ Digital Multitrack
- ○ DAW
- ○ Film

Other Resources
- ○ Professional Studios
- ○ Television Stations
- ○ Radio Stations
- ◉ Theater Tech Dept

Class Size Lecture *20* **Class Size Lab** *1*

Assistance
- ◉ Housing
- ◉ Scholarships
- ○ Internships
- ◉ Financial Aid
- ◉ Work-study
- ○ Job Placement

Admission Policy *Selective* **Language** *English*

Prerequisites
- ◉ High School Diploma
- ◉ SAT-ACT
- ○ Music Audition

Industry Affiliations
- ○ AES
- ○ ASA
- ○ NACB
- ○ NARAS
- ○ SMPTE
- ○ NAMBI
- ○ APRS
- ○ NAB
- ○ NAMM
- ○ SBE
- ○ SPARS
- ○ MEIEA

Fulltime Faculty *2* **Parttime Faculty**

Faculty Awards

Program Awards

Student Awards

Research areas

University of California at Santa Cruz

THE UCSC Electronic Music Program consists of a two-year sequence of courses offered by the Board of Studies in music. Electronic Music is not in itself a major; classes taken are part of a four-year undergraduate degree. The sequence begins with a general course on the history, theory, and literature of electronic music. Selected students are then admitted to a full-year series of hands-on studio courses which cover a variety of electronic music techniques, from tape splicing to SMPTE synchronization. Students then work on individual projects with various faculty members until they graduate.

Classes are small and flexible enough to accommodate a wide variety of technical and musical backgrounds. Students who complete the entire sequence of courses will be well-versed in synthesis, recording, MIDI, SMPTE, and computer techniques. Most of all, they will be able to utilize these skills to realize their own musical ideas.

Currently housed in three studios in the UCSC Communications Building, Electronic Music will acquire an expanded home with the completion of the school's new music building in 1995. Each studio contains a core of recording, signal processing, mixing, and monitoring gear. Synthesizers, ranging from classic modular analog to the latest MIDI instruments, are distributed according to class requirements. All equipment is installed for convenient and efficient use and is handicapped accessible. All gear is regularly maintained and calibrated by an in-house technician, and obsolete or worn out equipment is replace promptly.

Compositional computer systems are Macintosh-based, running the latest version of a variety of software. Students wishing to learn MIDI programming can use either Macintosh or PC platforms. SMPTE synchronization is available for video soundtrack work or automated mixing of multitrack recordings.

The electronic music courses are not limited to music majors. Musicians are joined by equal numbers of theater or film/video students. Other majors represented range from computer science to religious studies. Graduates and undergraduates work together in the same courses, all meeting the same rigorous criteria for sonic and creative quality of their productions. Each student is expected to devote at least eight hours a week to composition in the studios. Finished products are heard in studio and individual recitals or as part of dance, theater, or video productions.

Graduates of the program are found in all areas of the music industry, including recording studios, theater/film/video production, and high-tech music companies.

University of Colorado at Denver, Music Department

Address PO Box 173364, Campus Box 162
Denver CO
80217-3364 USA

Phone 303-556-2727 **FAX** 303-556-2335

Director Roy Pritts **Admission Contact** Holly Allen

Program Founded 1970 **School Type** University

Program Offered Music Engineering, Music Management

Degrees Offered Bachelor of Science

Program Length 4 years

Estimated Tuition $11,000

Main Emphasis Music Recording **Program is** Semi-technical

Accreditations NASM

Number of Studios 3 **Is school non-profit?** Yes

Types of Studios ○ Acoustic Research ◉ Electronic Music ○ Film/Foley ○ Radio
○ Audio Research ◉ Music Recording ○ Television ○ Video

Types of Recording ◉ Analog Multitrack ◉ MIDI Sequencing ◉ Video
○ Digital Multitrack ◉ DAW ○ Film

Other Resources ◉ Professional Studios ◉ Television Stations
◉ Radio Stations ◉ Theater Tech Dept

Class Size Lecture 15 **Class Size Lab** 3

Assistance ○ Housing ◉ Scholarships ◉ Internships
◉ Financial Aid ◉ Work-study ○ Job Placement

Admission Policy Selective **Language** English

Prerequisites ◉ High School Diploma ◉ SAT-ACT ◉ Music Audition

Industry Affiliations ◉ AES ○ ASA ○ NACB ◉ NARAS ◉ SMPTE ◉ NAMBI
◉ APRS ◉ NAB ◉ NAMM ○ SBE ◉ SPARS ◉ MEIEA

Fulltime Faculty 3 **Parttime Faculty** 4

Faculty Awards

Program Awards

Student Awards

Research areas

University of Colorado at Denver, Music Department

The music program at the University of Colorado at Denver prepares students for professional careers in the music industry. The first two years of study offer courses in the arts and sciences and include a firm background in music theory, literature, and performance. At the completion of the second year, students select one of four areas of study: music engineering, music management, performance, or scoring and arranging. The specialized curriculum has led graduates to local, regional, and national positions in audio research, production, musical arts administration, and audio engineering as well as graduate studies at leading universities and conservatories. Many graduates are also established musicians or owners of booking agencies, publishing companies, and recording studios.

The Music Engineering emphasis addresses contemporary technology in studio recording, sound reinforcement, and electronic music. It is intended to develop skills for creative musicians, producers, and technicians, using both analog and digital technology. The engineering studies stress artistic applications of technology to recording, reinforcement, composition, and performance. All of the studies in Music Engineering are available to non-degree students with the exception of certain areas of Applied Study and Independent Study. The school encourages working and parttime students.

The Music Engineering program operates two multitrack control rooms as well as three mixdown control rooms which are all interconnected to five music performance areas of varying size. The facilities are supported by regular gifts from the industry. The Music Department also features several electronic music and computer music labs.

The school is an institutional member of SPARS and participates in their internship program. The campus also hosts a student section of the AES. They regularly organize special events of interest to a broad audio population and attract students and professionals from throughout the Rocky Mountain area.

The Music Department also offers an emphasis in Music Management. The program is for students interested in preparing for careers in such fields as artist management, music publishing, music merchandising, concert promotion, record production, and telecommunications. The school is an affiliate of the National Association of Music Business Institutes.

University of Hartford, College of Engineering

Address	200 Bloomfield Avenue West Hartford 06117-0395	CT USA	
Phone	203-768-4792	**FAX**	203-768-5073
Director	Robert Celmer	**Admission Contact**	Robert Celmer
Program Founded	1976	**School Type**	University
Program Offered	Acoustics and Music		
Degrees Offered	Bachelor of Science in Engineering		
Program Length	4 years		
Estimated Tuition	$45,000		
Main Emphasis	Audio Engineering	**Program is**	Technical
Accreditations	State of Connecticut		
Number of Studios	6+	**Is school non-profit?**	Yes
Types of Studios	◉ Acoustic Research ◉ Electronic Music ○ Film/Foley ◉ Radio ◉ Audio Research ◉ Music Recording ◉ Television ○ Video		
Types of Recording	◉ Analog Multitrack ◉ MIDI Sequencing ○ Video ◉ Digital Multitrack ○ DAW ○ Film		
Other Resources	◉ Professional Studios ○ Television Stations ◉ Radio Stations ○ Theater Tech Dept		
Class Size Lecture	20	**Class Size Lab**	10
Assistance	◉ Housing ◉ Scholarships ◉ Internships ◉ Financial Aid ◉ Work-study ○ Job Placement		
Admission Policy	Selective	**Language**	English
Prerequisites	◉ High School Diploma ◉ SAT-ACT ◉ Music Audition		
Industry Affiliations	◉ AES ◉ ASA ○ NACB ○ NARAS ○ SMPTE ○ NAMBI ○ APRS ○ NAB ○ NAMM ○ SBE ○ SPARS ○ MEIEA		
Fulltime Faculty	15	**Parttime Faculty**	20
Faculty Awards	ASEE Young Faculty Award		
Program Awards			
Student Awards	NSF Research Semester		
Research areas	Noise Control, Musical Acoustics, Hearing Loss		

University of Hartford, College of Engineering

The University of Hartford offers a combined program in Acoustics and Music through their Interdisciplinary Engineering Studies Program within the College of Engineering. The four-year program results in a Bachelor of Science in Engineering degree and includes a basic engineering core as well as a major concentration in vibrations, acoustics, and music courses offered by the Hartt School of Music. The program is designed for those interested in audio engineering, noise control, architectural acoustics, musical acoustics, and bio-engineering.

The program was instituted in 1976 by Conrad Hemond, Jr. of the College of Engineering's Acoustics Lab, and William Willett of the Hartt School of Music. The degree is designed for students with an aptitude and desire for a career involving the application of technology to the field of music and/or acoustics. Applicants to the program must have the math and science background required of all engineering students and must pass the entrance requirements, including an audition, of the Hartt School of Music.

Although the curriculum is scheduled to be completed in four years, it is the most rigorous undergraduate program at the University, requiring a minimum of 142 credit hours for completion and often extending to 146 credits. Acoustics and Music students are typically the school's brightest students. Their dual pursuit brings a special perspective to their studies, and graduates have a personalized, career-oriented education with which they can use engineering as a broad base to pursue a variety of careers in acoustics and/or music.

The Acoustics Lab at the College of Engineering is a professional acoustics facility which serves the needs of students, faculty, and other professionals. Facilities include an anechoic sound chamber, acoustic and sound measuring devices, and related equipment. Students often work on real-world acoustic application projects coordinated with local industries. Students also utilize the Hartt School of Music's recording and electronic music studios.

University of Iowa, School of Music

Address *2057 Music Building*
Iowa City *Iowa*
52242-1793 *USA*

Phone *319-335-1664* **FAX** *319-335-2777*

Director *Lowell Cross* **Admission Contact** *Admissions Office*

Program Founded *1980* **School Type** *Workshop*

Program Offered *Seminar in Audio Recording*

Degrees Offered *None for Summer Workshop*

Program Length *Two weeks*

Estimated Tuition *$184-$290*

Main Emphasis *Music Recording* **Program is** *Technical*

Accreditations *NCACSS, AAU, CIC, NASM*

Number of Studios *2* **Is school non-profit?** *Yes*

Types of Studios ○ *Acoustic Research* ○ *Electronic Music* ○ *Film/Foley* ○ *Radio* ○ *Audio Research* ◉ *Music Recording* ○ *Television* ○ *Video*

Types of Recording ◉ *Analog Multitrack* ○ *MIDI Sequencing* ○ *Video* ○ *Digital Multitrack* ◉ *DAW* ○ *Film*

Other Resources ○ *Professional Studios* ○ *Television Stations* ○ *Radio Stations* ○ *Theater Tech Dept*

Class Size Lecture *20* **Class Size Lab** *Varies*

Assistance ◉ *Housing* ○ *Scholarships* ◉ *Internships* ○ *Financial Aid* ◉ *Work-study* ○ *Job Placement*

Admission Policy *Selective* **Language** *English*

Prerequisites ◉ *High School Diploma* ◉ *SAT-ACT* ◉ *Music Audition*

Industry Affiliations ◉ *AES* ○ *ASA* ○ *NACB* ○ *NARAS* ○ *SMPTE* ○ *NAMBI* ○ *APRS* ○ *NAB* ○ *NAMM* ○ *SBE* ○ *SPARS* ○ *MEIEA*

Fulltime Faculty *1* **Parttime Faculty** *Varies*

Faculty Awards

Program Awards

Student Awards

Research areas *Microphone Techniques*

University of Iowa, School of Music

Since 1980, the University of Iowa's School of Music has offered the Seminar in Audio Recording, an intensive two-week summer workshop. The program provides instruction in professional techniques of music recording. The goal is to reacquaint serious students with fundamental principles of audio, to explore the latest developments in studio techniques and technologies, and to provide a forum for experienced engineers and producers to discuss contemporary issues with their colleagues.

The seminar provides intensive training and learning opportunities. Four hours are spent daily in classroom and studio lectures, with equipment available for deomonstrations throughout the two-week period. Although the course is designed for persons familiar with recording equipment, studio technology, and recording techniques, beginning students are not excluded. Students are welcome to suggest discussion topics, while much of the agenda is determined by instructors and invited guests. Visiting instructors have included Bob Ludwig, John Eargle, Russell Hamm, Jerry Bruck, Jürgen Wahl, and Greg Calbi. The program is directed by Lowell Cross, professor of music and director of the University of Iowa Recording Studios.

Recordings at the School of Music are made in two professional control studios which are interconnected to the 2,600 seat Hancher Auditorium, the 720-seat Clapp Recital Hall, and ten other halls and studios in the Music Building. The facilities are equipped with state-of-the-art microphones, consoles, and multitrack recorders, as well as five different digital formats for recording and editing. Faculty and student performers provide audio students with recording opportunities with their musical performances.

Participants in past seminars have included established recording professionals, doctoral candidates from music schools, and students with career aspirations in music and engineering. The group represents all regions of the United States along with several other countries and has ranged in age from teenagers to those in their eighties. The seminar's enrollment is limited to 24 students.

Inexpensive lodging and meals are available from the University's Residence Services. Residence halls are within walking distance of the School of Music.

University of Massachusetts at Lowell

Address One University Avenue, College of Fine Arts
Lowell MA
01854 USA

Phone 508-934-3850 **FAX** 508-934-3034

Director William Moylan **Admission Contact** Admissions Office

Program Founded 1983 **School Type** University

Program Offered Sound Recording Technology

Degrees Offered Bachelor of Music in SRT, Minor in SRT for EE or Computer Science Majors

Program Length 4 years

Estimated Tuition $4,000 per year

Main Emphasis Music Recording **Program is** Semi-technical

Accreditations NASM

Number of Studios 5 **Is school non-profit?** Yes

Types of Studios ○ Acoustic Research ● Electronic Music ○ Film/Foley ○ Radio
○ Audio Research ● Music Recording ○ Television ● Video

Types of Recording ● Analog Multitrack ● MIDI Sequencing ● Video
○ Digital Multitrack ● DAW ○ Film

Other Resources ○ Professional Studios ○ Television Stations
○ Radio Stations ○ Theater Tech Dept

Class Size Lecture 17 **Class Size Lab** 8

Assistance ● Housing ● Scholarships ● Internships
● Financial Aid ● Work-study ● Job Placement

Admission Policy Selective **Language** English

Prerequisites ● High School Diploma ● SAT-ACT ● Music Audition

Industry Affiliations ● AES ○ ASA ○ NACB ● NARAS ○ SMPTE ○ NAMBI
○ APRS ○ NAB ○ NAMM ○ SBE ● SPARS ● MEIEA

Fulltime Faculty 1 **Parttime Faculty** 9

Faculty Awards

Program Awards 1992 TEC Nominee

Student Awards

Research areas

University of Massachusetts at Lowell

The University of Massachusetts at Lowell offers three degree programs in Sound Recording Technology: a Bachelor of Music degree, a Minor in SRT for Electrical Engineering Majors, and a Minor in SRT for Computer Science Majors. The University is the only institution in the country providing students with an opportunity to choose from three approaches to the industry: focusing on creative/production-oriented careers, on hardware development or facility design and maintenance, or software development. The programs are rigorous in their demands for quality performance from students, allowing nearly 90% of the programs' graduates to be placed throughout the audio industry.

The Bachelor of Music degree program is offered through the College of Fine Arts. The goal of the program is to produce musically sophisticated and sensitive professionals with sufficient technical knowledge to excel in the current industry and to easily keep pace with the changing technology. The program combines studies in physics, electrical engineering, computer science, and advanced mathematics with traditional studies in music. The program includes a minimum of nine courses in recording arts and technologies. The program culminates in either an internship or senior project.

The sequence of study for the Minor in SRT for Electrical Engineering Majors is offered in conjunction with the College of Engineering. It is designed to assist EE students interested in entering the recording industry as maintenance technicians or those interested in audio engineering research and development. The sequence of study focuses on providing students with practical knowledge of the function and use of of audio and video equipment, equipment maintenance, audio design theories and applications, and basic music skills. The student is introduced to artistic concepts and applications of audio and music synthesis and production while studying advanced concepts in audio theory. The Minor culminates with a research project in hardware design, including the building of a prototype of the design project.

The Minor in SRT for Computer Science Majors is aimed at those interested in audio-related software development and is offered in conjunction with the College of Pure and Applied Sciences, Department of Computer Science. The program's goal is to create technically articulate and knowledgeable individuals who share an awareness of musical concerns and concepts with the application of technology to musical processes.

All programs benefit from the University's state-of-the-art recording, mixing, sound reinforcement, video post-production, sound synthesis, and MIDI studios as well as equipment design and computer laboratories.

University of Miami- Electrical Engineering

Address *College of Engineering*
Coral Gables *FL*
33124-0640 *USA*

Phone *305-284-3291* **FAX** *305-284-4044*

Director *Reuven Lask* **Admission Contact** *Martina Hahn*

Program Founded *1992* **School Type** *University*

Program Offered *Audio Engineering*

Degrees Offered *Bachelor of Science in Electrical Engineering*

Program Length *4 years*

Estimated Tuition *$60,000*

Main Emphasis *Electrical Engineering* **Program is** *Technical*

Accreditations *ABET*

Number of Studios *2* **Is school non-profit?** *Yes*

Types of Studios
- ○ *Acoustic Research* ● *Electronic Music* ○ *Film/Foley* ● *Radio*
- ● *Audio Research* ● *Music Recording* ● *Television* ○ *Video*

Types of Recording
- ● *Analog Multitrack* ○ *MIDI Sequencing* ○ *Video*
- ○ *Digital Multitrack* ● *DAW* ○ *Film*

Other Resources
- ● *Professional Studios* ○ *Television Stations*
- ○ *Radio Stations* ○ *Theater Tech Dept*

Class Size Lecture *25* **Class Size Lab** *10*

Assistance
- ● *Housing* ● *Scholarships* ● *Internships*
- ● *Financial Aid* ● *Work-study* ● *Job Placement*

Admission Policy *Highly selective* **Language** *English*

Prerequisites ● *High School Diploma* ● *SAT-ACT* ○ *Music Audition*

Industry Affiliations
- ● *AES* ○ *ASA* ○ *NACB* ○ *NARAS* ● *SMPTE* ○ *NAMBI*
- ○ *APRS* ○ *NAB* ○ *NAMM* ○ *SBE* ○ *SPARS* ○ *MEIEA*

Fulltime Faculty *4* **Parttime Faculty** *0*

Faculty Awards *AES Fellow, AES Board of Governors Award*

Program Awards

Student Awards

Research areas *Digital Signal Processing*

University of Miami- Electrical Engineering

The University of Miami offers a Bachelor of Science in Electrical Engineering with Audio Engineering Option through the Department of Electrical and Computer Engineering. The program is the first degree of its kind in the United States to provide a BSEE with an emphasis in audio theory and practice. The programs combines traditional electrical engineering studies such as circuit theory, electronics, microprocessors, integrated circuits, and communication electronics, with audio studies in areas such as acoustics, digital audio, transducers, signal processing, post-production, and recording. Prerequisite courses in mathematics, physics, and chemistry are also included.

To create the curriculum, the Electrical and Computer Engineering department joined with the Music Engineering Technology program in the School of Music, an established program with a tradition of excellence. While the Music Engineering Technology program is aimed at students with strong musical backgrounds who wish to continue their studies in music, the Audio Engineering program is designed for students desiring a stronger technical emphasis, culminating in an electrical engineering degree. Graduates from the program may continue their studies in Audio Engineering by seeking admission to the School of Music's Master of Science degree in Audio Engineering.

Students enrolled in the Audio Engineering program have access to the same faculty and facilities as those in the Music Engineering Technology program. For more information about the faculty and facilities, as well as the other degree offerings from the school, see University of Miami- Music.

University of Miami- Music

Address	*School of Music* *Coral Gables* *33124-7610*	*FL* *USA*	
Phone	*305-284-2245*	**FAX**	*305-284-6475*
Director	*Ken Pohlmann*	**Admission Contact**	*Kenneth Moses*
Program Founded	*1975*	**School Type**	*University*
Program Offered	*Music Engineering Technology*		
Degrees Offered	*Bachelor of Music, Master of Science*		
Program Length	*4 years, 2 years*		
Estimated Tuition	*$60,000*		
Main Emphasis	*Audio Engineering*	**Program is**	*Technical*
Accreditations	*NASM*		
Number of Studios	*2*	**Is school non-profit?**	*Yes*
Types of Studios	○ *Acoustic Research* ◉ *Electronic Music* ○ *Film/Foley* ◉ *Radio* ◉ *Audio Research* ◉ *Music Recording* ◉ *Television* ○ *Video*		
Types of Recording	◉ *Analog Multitrack* ○ *MIDI Sequencing* ○ *Video* ○ *Digital Multitrack* ◉ *DAW* ○ *Film*		
Other Resources	◉ *Professional Studios* ○ *Television Stations* ○ *Radio Stations* ○ *Theater Tech Dept*		
Class Size Lecture	*25*	**Class Size Lab**	*10*
Assistance	◉ *Housing* ◉ *Scholarships* ◉ *Internships* ◉ *Financial Aid* ◉ *Work-study* ◉ *Job Placement*		
Admission Policy	*Highly selective*	**Language**	*English*
Prerequisites	◉ *High School Diploma* ◉ *SAT-ACT* ◉ *Music Audition* *BSEE, GRE for Master's Degree*		
Industry Affiliations	◉ *AES* ○ *ASA* ○ *NACB* ○ *NARAS* ◉ *SMPTE* ○ *NAMBI* ○ *APRS* ○ *NAB* ○ *NAMM* ○ *SBE* ○ *SPARS* ○ *MEIEA*		
Fulltime Faculty	*3*	**Parttime Faculty**	*0*
Faculty Awards	*AES Fellow, AES Board of Governors Award*		
Program Awards	*TEC Award*		
Student Awards	*Downbeat, Grammys, Emmys, etc.*		
Research areas	*Digital Signal Processing*		

University of Miami- Music

The University of Miami offers three options for earning degrees in Music Engineering Technology. The first is a four-year program culminating in a Bachelor of Music degree with a minor in electrical engineering. The objective of this program is to prepare musicians for professional careers as recording engineers, sound reinforcement engineers, audio technicians, manufacturers' representatives, and other positions in the music industry. Course work is designed to give a theoretical knowledge of audio sciences, with a special emphasis placed on digital audio technology. Extensive hands-on experience complements lecture material, providing a solid, practical basis of understanding. Graduates have enjoyed a high placement rate in the professional audio industry.

The Bachelor of Music program encompasses three main areas of study: music, audio engineering, and electrical engineering. Music courses include music theory, principal and secondary instrumental studies, music history and literature, and orchestration and arranging. Audio engineering courses include audio theory, recording techniques, digital audio, acoustics and studio design, audio for video, sound synthesis, and physics. Electrical engineering courses include computer programming, circuit theory, and logic design. Additional courses are required in calculus, video production, music business, and marketing. Internships are available for study in the professional industry.

For those looking for a graduate program, the Master of Science degree in Music Engineering Technology combines courses in Electrical Engineering and Music Engineering to enable advanced study in areas such as digital signal processing, psychoacoustics, and digital audio. The curriculum that of many graduate electrical engineering curricula, but includes courses emphasizing audio theory and practice. Digital signal processing is the recommended area of emphasis, but the curriculum is designed to be flexible enough to meet individual student's interests. Graduates from the program are prepared for positions in the audio industry such as audio software and hardware design engineers, studio circuit and system designers, and studio chief engineers.

The graduate degree requires courses in audio studies, electrical engineering, psychoacoustics, and a research project. The research project is part of the second year of study and culminates in a research paper and an oral defense. Applicants to the program must have completed a Bachelor of Science in Electrical Engineering. In addition, a background in audio engineering, an interest in music, and competitive GRE scores are expected.

Students enrolled in the Music Engineering Technology programs have access to two recording facilities. The Gusman Concert Hall contains a professional recording studio with an automated console and multitrack recorder and is used to record live concerts ranging from small jazz groups to large symphony orchestras. The Weeks Center for Recording and Performance is a new recording facility aimed at tracking and remix sessions. In addition, students have access to digital audio workstations, signal processing equipment, audio test gear, and a variety of computer platforms.

The program's faculty are widely known in the fields of music and audio engineering. They have published numerous articles, books, and scientific papers, chaired technical symposiums, and consulted in a variety of areas in the professional industry, from recording studio design to automobile sound system design.

The third degree program offered at the University of Miami is a Bachelor of Science in Electrical Engineering with an emphasis in Audio Engineering and is detailed elsewhere in the handbook. See University of Miami- Engineering.

University of New Haven

Address	*300 Orange Avenue*	
	West Haven	*CT*
	06516	*USA*
Phone	*203-932-7101*	**FAX**
Director	*Michael Kaloyanides*	**Admission Contact** *Michael Kaloyanides*
Program Founded	*1985*	**School Type** *University*
Program Offered	*Music and Sound Recording, Music Industry*	
Degrees Offered	*Bachelor of Arts, Bachelor of Science*	
Program Length	*4 years*	
Estimated Tuition	*$40,000*	
Main Emphasis	*Music Recording*	**Program is** *Technical*
Accreditations	*NEASC*	
Number of Studios	*2*	**Is school non-profit?** *Yes*
Types of Studios	○ *Acoustic Research* ○ *Electronic Music* ○ *Film/Foley* ○ *Radio* ○ *Audio Research* ◉ *Music Recording* ○ *Television* ○ *Video*	
Types of Recording	◉ *Analog Multitrack* ◉ *MIDI Sequencing* ○ *Video* ○ *Digital Multitrack* ◉ *DAW* ○ *Film*	
Other Resources	○ *Professional Studios* ○ *Television Stations* ◉ *Radio Stations* ◉ *Theater Tech Dept*	
Class Size Lecture	*15*	**Class Size Lab** *4*
Assistance	○ *Housing* ○ *Scholarships* ◉ *Internships* ◉ *Financial Aid* ◉ *Work-study* ◉ *Job Placement*	
Admission Policy	*Selective*	**Language** *English*
Prerequisites	◉ *High School Diploma* ◉ *SAT-ACT* ○ *Music Audition*	
Industry Affiliations	◉ *AES* ○ *ASA* ○ *NACB* ○ *NARAS* ○ *SMPTE* ○ *NAMBI* ○ *APRS* ○ *NAB* ○ *NAMM* ○ *SBE* ○ *SPARS* ◉ *MEIEA*	
Fulltime Faculty	*4*	**Parttime Faculty** *6*
Faculty Awards		
Program Awards		
Student Awards		
Research areas		

University of New Haven

The University of New Haven offers three degree programs related to recording and the music industry: a Bachelor of Arts in Music and Sound Recording, a Bachelor of Science in Music and Sound Recording, and a Bachelor of Arts in Music Industry. The programs are designed to meet the needs of today's musicians. In addition to providing students with a strong traditional background in music, the degree programs supply students with the knowledge and skills needed to enter the professional world or to pursue graduate study.

The Bachelor of Arts in Music and Sound Recording is based on the philosophy that musicians should have a working knowledge of the music-related media and that sound recordists should have a working knowledge of music. The program is designed to instruct students in three areas: music history, theory and aesthetics; musicianship; and sound recording methodology and technique. Courses are required in arts and science, music, recording, and elective studies.

The Bachelor of Science in Music and Sound Recording is similar to the Bachelor of Arts program in its philosophy and design, but provides a stronger background in the science and technology of recording through classes in calculus, physics, and electrical engineering in addition to the required courses in arts and science, music, and recording.

The Bachelor of Music Industry provides a balance of courses in the areas of music, sound recording, and business as well as music industry. The music industry courses cover topics such as record companies, contracts, music marketing and merchandising, recording studio management, music publishing, copyright law, event production, promotion, and management. Special emphasis is given to career planning and development. Courses are required in arts and science, music and sound recording, music industry and business, and elective studies.

A new multitrack recording facility is utilized for all recording technology classes and provides students with a professional recording environment. The control room was designed to allow comfortable seating for students while providing an excellent view of the console and related equipment. A second multitrack facility is also available as well as several recording/mixdown stations.

Students may participate in a special coop program which combines fulltime study with paid experience in their field. Coop students alternate semesters of employment with course work, resulting in a five-year program. Students can also utilizes internships to gain professional experience.

University of North Alabama

Address	*Box 5040*		
	Florence	*AL*	
	35632-0001	*USA*	
Phone	*205-760-4361*	**FAX**	*205-760-4329*
Director	*James Simpson*	**Admission Contact**	*Fred Alexander*
Program Founded	*1975*	**School Type**	*University*
Program Offered	*Commercial Music*		
Degrees Offered	*Bachelor of Arts/Science*		
Program Length	*4 years*		
Estimated Tuition	*$18,000*		
Main Emphasis	*Music Business*	**Program is**	*Non-technical*
Accreditations			
Number of Studios	*1*	**Is school non-profit?**	*Yes*

Types of Studios
- O *Acoustic Research*
- O *Electronic Music*
- O *Film/Foley*
- O *Radio*
- O *Audio Research*
- ◉ *Music Recording*
- O *Television*
- O *Video*

Types of Recording
- ◉ *Analog Multitrack*
- O *MIDI Sequencing*
- O *Video*
- ◉ *Digital Multitrack*
- O *DAW*
- O *Film*

Other Resources
- O *Professional Studios*
- O *Television Stations*
- O *Radio Stations*
- O *Theater Tech Dept*

Class Size Lecture *15* **Class Size Lab** *15*

Assistance
- O *Housing*
- O *Scholarships*
- O *Internships*
- ◉ *Financial Aid*
- ◉ *Work-study*
- O *Job Placement*

Admission Policy *Open* **Language** *English*

Prerequisites
- ◉ *High School Diploma*
- O *SAT-ACT*
- O *Music Audition*

Industry Affiliations
- O *AES*
- O *ASA*
- O *NACB*
- O *NARAS*
- O *SMPTE*
- O *NAMBI*
- O *APRS*
- O *NAB*
- O *NAMM*
- O *SBE*
- O *SPARS*
- O *MEIEA*

Fulltime Faculty *0* **Parttime Faculty** *3*

Faculty Awards

Program Awards

Student Awards

Research areas

University of North Alabama

The Commercial Music program at the University of North Alabama places students in the midst of the business and creative side of the commercial music industry, affording them the opportunity to study the various segments of the industry. The primary objective of the program is to prepare students with enough knowledge and classroom experience to enter the commercial music field. Since the University is just across the river from the Muscle Shoals music industry, students benefit from a regular schedule of guest speakers from the music industry.

The program offers the option of pursuing either a Bachelor of Arts in Music or a Bachelor of Science in Music. Required courses are balanced between general studies, music, and business areas. Specific commercial music courses include: Survey of the Music Industry, Popular Songwriting, Music Publishing, The Record Company, Studio Techniques, Music Production, and Commercial Music Paractium. The Commercial Music Practicum is an internship in a professional area such as publishing, engineering, record merchandising, or radio. The program has placed students in interns in Nashville and Los Angeles in addition to the local music industry.

University of North Carolina at Asheville

Address *UNCA Music Department*
Asheville *NC*
28804-3299 *USA*

Phone *704-258-6432* **FAX** *704-251-6385*

Director *Wayne Kirby* **Admission Contact** *Wayne Kirby*

Program Founded *1982* **School Type** *University*

Program Offered *Music with Recording Arts*

Degrees Offered *Bachelor of Science*

Program Length *4 years*

Estimated Tuition *$640-$3,097 semester*

Main Emphasis *Music Recording* **Program is** *Technical*

Accreditations *NASM*

Number of Studios *6* **Is school non-profit?** *Yes*

Types of Studios ○ *Acoustic Research* ◉ *Electronic Music* ○ *Film/Foley* ○ *Radio*
○ *Audio Research* ◉ *Music Recording* ○ *Television* ○ *Video*

Types of Recording ○ *Analog Multitrack* ○ *MIDI Sequencing* ○ *Video*
○ *Digital Multitrack* ○ *DAW* ○ *Film*

Other Resources ◉ *Professional Studios* ○ *Television Stations*
○ *Radio Stations* ◉ *Theater Tech Dept*

Class Size Lecture *Varies* **Class Size Lab** *Varies*

Assistance ◉ *Housing* ◉ *Scholarships* ◉ *Internships*
◉ *Financial Aid* ◉ *Work-study* ◉ *Job Placement*

Admission Policy *Selective* **Language** *English*

Prerequisites ◉ *High School Diploma* ◉ *SAT-ACT* ◉ *Music Audition*

Industry Affiliations ◉ *AES* ○ *ASA* ○ *NACB* ○ *NARAS* ○ *SMPTE* ○ *NAMBI*
○ *APRS* ○ *NAB* ○ *NAMM* ○ *SBE* ○ *SPARS* ○ *MEIEA*

Fulltime Faculty *4* **Parttime Faculty** *18*

Faculty Awards

Program Awards

Student Awards

Research areas

University of North Carolina at Asheville

The University of North Carolina at Asheville offers a Bachelor of Science in Music with Recording Arts program for students interested in both the creative and technological aspects of music recording and production. To further specialize, students may minor in computer science, management, mass communications, or theater. Non-music students who are interested in preparing for the music or entertainment industries may minor in Music with an emphasis in Music Business or Recording Arts. It is also possible to pursue individual degree programs which are outside of the normal list of major programs and combine features of more than one major.

The Music with Recording Arts program was established in 1982. It is a small and highly selective program taught by a distinguished faculty that includes Dr. Robert Moog, inventor of the Moog synthesizer. The program is grounded in a liberal arts and music curriculum that prepares students for entry-level positions as well as leadership roles in the music professions of the next century.

Music technology topics covered by the program include: audio engineering technology, MIDI, analog and digital synthesis, electronic music, audio electronics, directed studio projects, and music business. Qualified students may earn up to six credits while participating in an internship with a participating firm. Many interns are placed in major entertainment industry centers such as New York, Nashville, Los Angeles, and elsewhere.

The Recording Arts program benefits from the Department of Music's Aural Skills Lab, Electronic Keyboard Lab, Computer Music Lab, and three UNCA Recording Studios. The recording facilities consist of one on-campus, multitrack recording studio and two off-campus, professional recording studios which are contracted for the use of advanced Recording Arts students. Sound reinforcement systems and audio testing and measurement devices are also available.

In addition to being accepted to UNCA, students wishing admittance to the Recording Arts program must pass a music performance audition, complete Class Piano II or demonstrate equivalent competency, and either pass the UNCA Music Theory/Aural Skills Entrance Examination or successfully complete Introductory Music Theory. Auditions and placement tests are arranged on an individual basis by calling or writing the UNCA Department of Music. In some cases, a cassette tape may be submitted in lieu of an in-person audition.

University Of South Florida

Address	*School of Mass Communications*		
	Tampa	*FL*	
	33620-7800	*USA*	
Phone	*813-974-2851*	**FAX**	
Director	*Donn Dickerson*	**Admission Contact**	*Admissions:*
Program Founded	*1956*	**School Type**	*University*
Program Offered	*Mass Communications*		
Degrees Offered	*Bachelor of Arts, Master of Arts*		
Program Length	*4 years, 2 years*		
Estimated Tuition	*$5,100-$18,240*		
Main Emphasis	*Broadcasting*	**Program is**	*Semi-technical*
Accreditations	*Various*		
Number of Studios	*8*	**Is school non-profit?**	*Yes*

Types of Studios
- ○ *Acoustic Research* ○ *Electronic Music* ○ *Film/Foley* ○ *Radio*
- ○ *Audio Research* ○ *Music Recording* ○ *Television* ○ *Video*

Types of Recording
- ◉ *Analog Multitrack* ○ *MIDI Sequencing* ◉ *Video*
- ○ *Digital Multitrack* ◉ *DAW* ○ *Film*

Other Resources
- ◉ *Professional Studios* ◉ *Television Stations*
- ◉ *Radio Stations* ◉ *Theater Tech Dept*

Class Size Lecture	*20*	**Class Size Lab**	*15*

Assistance
- ◉ *Housing* ◉ *Scholarships* ◉ *Internships*
- ◉ *Financial Aid* ◉ *Work-study* ○ *Job Placement*

Admission Policy	*Selective*	**Language**	*English*

Prerequisites
- ◉ *High School Diploma* ◉ *SAT-ACT* ○ *Music Audition*

Industry Affiliations
- ◉ *AES* ○ *ASA* ○ *NACB* ○ *NARAS* ○ *SMPTE* ○ *NAMBI*
- ○ *APRS* ◉ *NAB* ○ *NAMM* ○ *SBE* ○ *SPARS* ○ *MEIEA*

Fulltime Faculty	*15*	**Parttime Faculty**	*6*

Faculty Awards

Program Awards

Student Awards *Hurst Journalism Award*

Research areas *Mass Media Effects, Freedom of Information*

University Of South Florida

The University of South Florida offers undergraduate and graduate degrees in Mass Communications with an emphasis in audio production. The Mass Communications department offers approximately 70 courses varying in content from highly technical and field-specialized areas to essential liberal arts areas. The program introduces students to the theories, principles, and problems of communications, emphasizing the concept of freedom of information and preparing students for future leadership roles in communications media. Majors seeking careers in mass media will be directed to the various media with which the department maintains close contact for summer internships and part-time work. A limited number of students have opportunities to serve as interns with a mass communications organization and receive academic credit.

The Bachelor of Arts curriculum includes courses in audio production in media, advanced audio production, television production, announcing, broadcast writing, broadcast news, electronic news gathering, and various special topics offered over the summer. The program is also in the process of forming a student chapter of the AES.

The program moved into a new building in 1991 and benefits from three studios available for audio production work. The main facility is a live performance studio for music recording. A multitrack production studio with signal processing and digital recording is also available as well as a Macintosh-based digital audio workstation studio for MIDI and other projects.

University of Southern California

Address	*School of Music, MUS 409*		
	Los Angeles	*CA*	
	90089-0851	*USA*	
Phone	*213-740-3217*	**FAX**	*213-740-3217*
Director	*Richard McIlvery*	**Admission Contact**	*Richard McIlvery*
Program Founded	*1975*	**School Type**	*University*
Program Offered	*Music Recording*		
Degrees Offered	*Bachelor of Science*		
Program Length	*4 years*		
Estimated Tuition	*$60,000*		
Main Emphasis	*Audio Engineering*	**Program is**	*Semi-technical*
Accreditations	*NASM*		
Number of Studios	*2*	**Is school non-profit?**	*Yes*

Types of Studios
○ *Acoustic Research* ◉ *Electronic Music* ○ *Film/Foley* ○ *Radio*
○ *Audio Research* ◉ *Music Recording* ○ *Television* ○ *Video*

Types of Recording
◉ *Analog Multitrack* ◉ *MIDI Sequencing* ○ *Video*
◉ *Digital Multitrack* ◉ *DAW* ○ *Film*

Other Resources
◉ *Professional Studios* ○ *Television Stations*
○ *Radio Stations* ○ *Theater Tech Dept*

Class Size Lecture *19* **Class Size Lab** *12*

Assistance
◉ *Housing* ○ *Scholarships* ◉ *Internships*
◉ *Financial Aid* ◉ *Work-study* ○ *Job Placement*

Admission Policy *Highly selective* **Language** *English*

Prerequisites
◉ *High School Diploma* ◉ *SAT-ACT* ◉ *Music Audition*

Industry Affiliations
◉ *AES* ○ *ASA* ○ *NACB* ◉ *NARAS* ○ *SMPTE* ○ *NAMBI*
○ *APRS* ○ *NAB* ○ *NAMM* ○ *SBE* ○ *SPARS* ○ *MEIEA*

Fulltime Faculty *2* **Parttime Faculty** *4*

Faculty Awards

Program Awards

Student Awards

Research areas

University of Southern California

The School of Music at the University of Southern California (USC) offers a variety of programs for those interested in entering the professional audio or commercial music business fields. The major program is a Bachelor of Science in Music Recording. In addition, a Minor in Music Recording is offered primarily for students who are pursuing a Bachelor of Electrical Engineering, and a Certificate in Music Recording is offered for those already possessing a Bachelors degree. All recording classes are taught by industry professionals in USC's two state-of-the-art recording studio.

The curriculum for the Bachelor of Science in Music Recording is balanced between courses in arts and sciences, music, and recording arts. The requirements for the Minor and Certificate basically consist of recording arts courses. USC has some of the finest university recording facilities in the country. The School of Music's main studio control room is designed as both a classroom and a state-of-the-art recording studio. The facility is tied to two large performing areas via video lines. The program also benefits from the School of Music's Electronic Music Studio and the School of Cinema's complete scoring facility. Advanced student have opportunities to record film scores for joint Music and Cinema projects. USC also offers an innovative program in Entertainment Engineering offered through the School of Cinema and the School of Engineering.

The School of Music is among the top ten music schools in the United States and has a student body of approximately 720 students. The faculty are drawn from every discipline within music and many are actively involved in the Los Angeles music and entertainment communities. Faculty in the Recording Arts Program are selected because of their unique abilities in recording, their active involvement in the professional audio industry in Los Angeles, and their ability to relate to students.

University of Surrey

Address *Department of Music*
Guilford *Surrey*
GU2 5XH *UK*

Phone *44-483-509317* **FAX** *44-483-509317*

Director *Dave Fisher* **Admission Contact** *Dave Fisher*

Program Founded **School Type** *University*

Program Offered *Music and Sound Recording*

Degrees Offered *Bachelor of Music*

Program Length *4 years*

Estimated Tuition *£12,000-£28,000*

Main Emphasis *Music Recording* **Program is** *Semi-technical*

Accreditations

Number of Studios *7* **Is school non-profit?** *Yes*

Types of Studios
○ *Acoustic Research* ● *Electronic Music* ○ *Film/Foley* ○ *Radio*
○ *Audio Research* ● *Music Recording* ○ *Television* ○ *Video*

Types of Recording
● *Analog Multitrack* ● *MIDI Sequencing* ○ *Video*
● *Digital Multitrack* ● *DAW* ○ *Film*

Other Resources
○ *Professional Studios* ○ *Television Stations*
○ *Radio Stations* ○ *Theater Tech Dept*

Class Size Lecture *Varies* **Class Size Lab** *Varies*

Assistance
● *Housing* ● *Scholarships* ● *Internships*
● *Financial Aid* ○ *Work-study* ● *Job Placement*

Admission Policy *Selective* **Language** *English*

Prerequisites
● *High School Diploma* ○ *SAT-ACT* ● *Music Audition*
Contact School for Others

Industry Affiliations
● *AES* ○ *ASA* ○ *NACB* ○ *NARAS* ○ *SMPTE* ○ *NAMBI*
● *APRS* ○ *NAB* ○ *NAMM* ○ *SBE* ○ *SPARS* ○ *MEIEA*

Fulltime Faculty *8* **Parttime Faculty**

Faculty Awards

Program Awards

Student Awards

Research areas *New Audio Technologies*

University of Surrey

The Department of Music at the University of Surrey offers a Bachelor of Music degree in Music and Sound Recording structured as a Tonmeister program. The curriculum combines the teaching of recording techniques and the scientific principles of audio engineering with selected musical studies. The four-year program includes a unique Professional Training Period during the third year of studies that allows students to work in selected recording, broadcasting, or other professional sound establishments. This gives students an opportunity to develop their practical skills alongside working professionals and to relate their university studies to the music and recording industries.

The first year of the program provides students with a solid foundation in the principles of sound recording and music. Courses cover recording techniques, electronics, knowledge of instruments, harmony and keyboard skills, acoustics, mathematics, and more. The second year adds courses in music technology, electro-acoustics, microprocessor applications. The third year is the Professional Training Period. The final year continues with studies in recording techniques, recording seminars, and electro-acoustics. The program culminates with a substantial written project and a portfolio of recordings.

Extensive opportunities are provided for performing and recording a wide range of music. These range from weekly lunch-hour concerts to numerous choral and orchestral concerts. The school also hosts concerts by touring choirs, orchestras, folk groups, and rock bands. All departmental concerts are recorded by a team of Tonmeisters, and cassettes are made available to students.

The program is housed in the Performing Arts Technology Studios (PATS). Completed in 1988, the building's facilities include thirteen practice rooms, two well-equipped recording studios and control rooms, three editing rooms, and electronic music studios. The school also has its own van for transporting equipment to remote recording sites. The University Library has a well-stocked audio room with scores, records, cassettes, CDs and playback equipment.

The program hosts a Tonmeister Open Day each June for interested students, and there are several other University Open Days throughout the year. Applications from mature and overseas students, who may not have the normal entry requirements, are welcomed and will be considered on their individual merits. The school also offers opportunities for graduate study and research in music recording.

University of Tennessee, Music Department

Address *1741 Volunteer Blvd.*
Knoxville *TN*
37996-2600 *USA*

Phone *615-974-7552* **FAX**

Director *Kenneth Jacobs* **Admission Contact** *Kenneth Jacobs*

Program Founded *1974* **School Type** *University*

Program Offered *Electronic Music Composition*

Degrees Offered *Bachelor of Arts/Music, Master of Music*

Program Length *4 years, 2 years*

Estimated Tuition *$2,000-5,300 per year*

Main Emphasis *Electronic Music* **Program is** *Semi-technical*

Accreditations *NASM*

Number of Studios *3* **Is school non-profit?** *Yes*

Types of Studios
- ○ *Acoustic Research* ◉ *Electronic Music* ○ *Film/Foley* ○ *Radio*
- ○ *Audio Research* ○ *Music Recording* ○ *Television* ○ *Video*

Types of Recording
- ◉ *Analog Multitrack* ◉ *MIDI Sequencing* ○ *Video*
- ○ *Digital Multitrack* ◉ *DAW* ○ *Film*

Other Resources
- ○ *Professional Studios* ○ *Television Stations*
- ○ *Radio Stations* ○ *Theater Tech Dept*

Class Size Lecture *20* **Class Size Lab** *1*

Assistance
- ◉ *Housing* ◉ *Scholarships* ○ *Internships*
- ◉ *Financial Aid* ◉ *Work-study* ○ *Job Placement*

Admission Policy *Open* **Language** *English*

Prerequisites ◉ *High School Diploma* ◉ *SAT-ACT* ◉ *Music Audition*

Industry Affiliations
- ○ *AES* ○ *ASA* ○ *NACB* ○ *NARAS* ○ *SMPTE* ○ *NAMBI*
- ○ *APRS* ○ *NAB* ○ *NAMM* ○ *SBE* ○ *SPARS* ○ *MEIEA*

Fulltime Faculty *3* **Parttime Faculty** *0*

Faculty Awards

Program Awards

Student Awards

Research areas

University of Tennessee, Music Department

The University of Tennessee's School of Music offers both undergraduate and graduate degree programs in Music Theory/Composition with an emphasis in Electronic Music. In addition to courses in general studies and music, students are required to study sound recording techniques, sound synthesis techniques, composition with electronic media, and computer music transcription.

Concerts and radio programs of student and faculty work are presented regularly, and students are encouraged to collaborate with other students in art, dance, and theater. In addition, students and faculty are represented frequently at international electronic music conferences and festivals.

University of Wisconsin- Oshkosh

Address *Music Department*
Oshkosh *WI*
54901 *USA*

Phone *414-424-4224* **FAX**

Director *Wally Messner* **Admission Contact** *James Kohn*

Program Founded *1982* **School Type** *University*

Program Offered *Recording Technology*

Degrees Offered *Bachelor of Music*

Program Length *4 years*

Estimated Tuition *$8,000*

Main Emphasis *Music Recording* **Program is** *Semi-technical*

Accreditations *NASM*

Number of Studios *2* **Is school non-profit?** *Yes*

Types of Studios ○ *Acoustic Research* ◉ *Electronic Music* ○ *Film/Foley* ◉ *Radio* ○ *Audio Research* ◉ *Music Recording* ◉ *Television* ◉ *Video*

Types of Recording ◉ *Analog Multitrack* ◉ *MIDI Sequencing* ○ *Video* ○ *Digital Multitrack* ○ *DAW* ○ *Film*

Other Resources ◉ *Professional Studios* ○ *Television Stations* ○ *Radio Stations* ○ *Theater Tech Dept*

Class Size Lecture *20* **Class Size Lab** *5-10*

Assistance ○ *Housing* ◉ *Scholarships* ◉ *Internships* ◉ *Financial Aid* ◉ *Work-study* ◉ *Job Placement*

Admission Policy *Selective* **Language** *English*

Prerequisites ◉ *High School Diploma* ◉ *SAT-ACT* ◉ *Music Audition*

Industry Affiliations ○ *AES* ○ *ASA* ○ *NACB* ○ *NARAS* ○ *SMPTE* ◉ *NAMBI* ○ *APRS* ○ *NAB* ○ *NAMM* ○ *SBE* ○ *SPARS* ○ *MEIEA*

Fulltime Faculty *20* **Parttime Faculty** *12*

Faculty Awards

Program Awards

Student Awards *NARAS Student Music Awards*

Research areas

University of Wisconsin- Oshkosh

The University of Wisconsin- Oshkosh offers a Recording Technology emphasis as part of its Bachelor of Music in Music Merchandising degree. The program's comprehensive curriculum consists of a 52 credit core of music courses, 19 credits of recording courses, 15 credits of business courses, and 44 credits of general education courses. The four-year program covers all aspects of recording production and business.

The Recording Technology sequence begins with classes in audio theory and recording principles. Subsequent courses provide students with increasing amounts of hands-on expereince in the studios. The recording curriculum culminates in a two-semester practicum course where students are responsible for entire recording projects, from start to finish. Upon completion of all course work, students spend their final semester interning at professional recording facilities.

The program offers multitrack recording facilities with automation, sequencing, time code equipment, and a variety of outboard gear. Students from the program have won two of ten NARAS Student Music Awards given in 1990 and 1991. Early application to the University is advised.

Weaver Education Center

Address 300 S. Spring Street
Greensboro NC
27401 USA

Phone 919-370-8282 **FAX** 919-370-8287

Director Ronald Follas **Admission Contact** Ronald Follas

Program Founded 1978 **School Type** High School

Program Offered Electronic Music

Degrees Offered High School Program

Program Length 2 years

Estimated Tuition None

Main Emphasis Electronic Music **Program is** Semi-technical

Accreditations SASC

Number of Studios 2 **Is school non-profit?** Yes

Types of Studios
○ Acoustic Research ● Electronic Music ● Film/Foley ○ Radio
○ Audio Research ○ Music Recording ○ Television ○ Video

Types of Recording
● Analog Multitrack ● MIDI Sequencing ○ Video
○ Digital Multitrack ○ DAW ○ Film

Other Resources
○ Professional Studios ○ Television Stations
○ Radio Stations ○ Theater Tech Dept

Class Size Lecture 11 **Class Size Lab** 6

Assistance
○ Housing ○ Scholarships ○ Internships
○ Financial Aid ○ Work-study ○ Job Placement

Admission Policy Open **Language** English

Prerequisites
○ High School Diploma ○ SAT-ACT ○ Music Audition
High School Juniors & Seniors

Industry Affiliations
○ AES ○ ASA ○ NACB ○ NARAS ○ SMPTE ○ NAMBI
○ APRS ○ NAB ○ NAMM ○ SBE ○ SPARS ○ MEIEA

Fulltime Faculty 1 **Parttime Faculty** 0

Faculty Awards

Program Awards

Student Awards

Research areas

Weaver Education Center

The Philip J. Weaver Educational Center is a centrally-located extension of the public high schools in Greensboro, North Carolina. Courses offered at the Center are unique to the system and are provided to enable students from the entire school system to benefit. The curriculum includes vocational, performing and visual arts, and advanced academic programs.

Students fortunate enough to live in the Greensboro Public School District can take advantage of an electronic music program offered at the Weaver Education Center. The program covers a variety of topics, including tape techniques, sound basics, electronic music composition and analysis, analog and digital synthesizer programming, sampling, recording studio techniques, and music for film. The program is supported by seven electronic music facilities equipped with a variety of analog and digital synthesizers and recording gear.

Classes are presented in a lecture/demonstration format and are of a technical nature. The aim of the program is to encourage students to get the most out of their own work and experience with electronic music and recording. Students work on such projects as tape loops, audio system design, small group composition, and individual electronic music compositions. Participation is emphasized, and students meet for lecture/demonstrations for three periods per week, labs for one period, and readings for one period.

Students from the Electronic Music program often continue their studies at recording workshops, audio engineering schools, and music schools. Some former students are now studio musicians and recording engineers in Nashville, New York, and Los Angeles.

Webster University

Address	*470 East Lockwood Avenue* *St Louis* *MO* *63119* *USA*
Phone	*314-968-6924* **FAX** *314-968-7077*
Director	*Barry Hufker* **Admission Contact** *Charles Beech*
Program Founded	*1989* **School Type** *University*
Program Offered	*Media Communications with Audio Production Emphasis*
Degrees Offered	*Bachelor of Arts*
Program Length	*4 years*
Estimated Tuition	*$32,000-$36,000*
Main Emphasis	*Audio Engineering* **Program is** *Semi-technical*
Accreditations	*Numerous*
Number of Studios	*4* **Is school non-profit?** *Yes*
Types of Studios	○ *Acoustic Research* ◉ *Electronic Music* ◉ *Film/Foley* ◉ *Radio* ○ *Audio Research* ◉ *Music Recording* ◉ *Television* ◉ *Video*
Types of Recording	◉ *Analog Multitrack* ◉ *MIDI Sequencing* ◉ *Video* ○ *Digital Multitrack* ◉ *DAW* ◉ *Film*
Other Resources	◉ *Professional Studios* ◉ *Television Stations* ◉ *Radio Stations* ◉ *Theater Tech Dept*
Class Size Lecture	*15* **Class Size Lab** *15*
Assistance	◉ *Housing* ◉ *Scholarships* ◉ *Internships* ◉ *Financial Aid* ◉ *Work-study* ◉ *Job Placement*
Admission Policy	*Selective* **Language** *English*
Prerequisites	◉ *High School Diploma* ◉ *SAT-ACT* ○ *Music Audition*
Industry Affiliations	◉ *AES* ○ *ASA* ◉ *NACB* ○ *NARAS* ○ *SMPTE* ○ *NAMBI* ○ *APRS* ○ *NAB* ○ *NAMM* ○ *SBE* ◉ *SPARS* ○ *MEIEA*
Fulltime Faculty	*2* **Parttime Faculty** *6*
Faculty Awards	
Program Awards	
Student Awards	
Research areas	

Webster University

Webster University offers a four-year program for students interested in music recording, audio for video/film, sound design for theater, multimedia, and radio/television broadcasting. A broadly based program, the school seeks to create audio engineers capable of working throughout the audio industry. The undergraduate program is flexible enough for individuals to pursue personal career goals within the context of a liberal arts curriculum. Hands-on production is emphasized. Students completing the program and passing faculty-judged portfolio reviews receive a Bachelor of Arts degree in Media Communications with an Audio Engineering Emphasis.

Audio classes meet and assignments are completed in the University's three audio studios. These include a multitrack recording studio, a production studio, and a radio broadcast studio for KSLH-FM, which broadcasts to St. Louis, St. Louis County, and parts of southern Illinois. Students may also study at Webster's four European campuses.

The audio faculty is comprised of working professionals from the audio industry, the most notable being Bill Porter, former recording engineer for Elvis Presley, Roy Orbison, the Everly Brothers, and the first inductee to the Audio Hall of Fame. The program also has an active series of guest speakers and invites audio professional from around the country to speaker on campus. Webster University's program is also active with the St. Louis chapter of the AES and is an associate member of SPARS, NACB, and the Missouri Association of Educational Broadcasters.

William Paterson College of New Jersey

Address *Music Department*
Wayne *NJ*
07470 *USA*

Phone *201-595-2315* **FAX** *201-595-2460*

Director *Stephen Marcone* **Admission Contact** *Dr. Woodworth*

Program Founded *1991* **School Type** *University*

Program Offered *Audio Recording*

Degrees Offered *Bachelor of Arts in Music with Audio Recording Emphasis*

Program Length *4 years*

Estimated Tuition *$4,000*

Main Emphasis *Audio Engineering* **Program is** *Semi-technical*

Accreditations *Middle States, NASM*

Number of Studios *2* **Is school non-profit?** *Yes*

Types of Studios ○ *Acoustic Research* ◉ *Electronic Music* ◉ *Film/Foley* ◉ *Radio*
○ *Audio Research* ◉ *Music Recording* ◉ *Television* ◉ *Video*

Types of Recording ◉ *Analog Multitrack* ◉ *MIDI Sequencing* ○ *Video*
○ *Digital Multitrack* ○ *DAW* ○ *Film*

Other Resources ○ *Professional Studios* ○ *Television Stations*
○ *Radio Stations* ○ *Theater Tech Dept*

Class Size Lecture *Varies* **Class Size Lab** *Varies*

Assistance ◉ *Housing* ◉ *Scholarships* ◉ *Internships*
◉ *Financial Aid* ◉ *Work-study* ◉ *Job Placement*

Admission Policy *Selective* **Language** *English*

Prerequisites ◉ *High School Diploma* ◉ *SAT-ACT* ○ *Music Audition*

Industry Affiliations ○ *AES* ○ *ASA* ○ *NACB* ○ *NARAS* ○ *SMPTE* ○ *NAMBI*
○ *APRS* ○ *NAB* ○ *NAMM* ○ *SBE* ○ *SPARS* ○ *MEIEA*

Fulltime Faculty *1* **Parttime Faculty** *8*

Faculty Awards

Program Awards

Student Awards

Research areas

William Paterson College of New Jersey

The William Paterson College of New Jersey offers a four-year Bachelor of Arts in Music Studies with a Concentration in Audio Recording. The program's curriculum is comprised of courses in music, audio recording, electronic music, and general studies. The curriculum has enough free electives to allow the student to specialize further in a variety of areas. Students benefit from a professionally-equipped recording studio as well as an extensive eight-station electronic music studio.

The school is conveniently located for students wishing to take advantage of internships and music industry opportunities in the New York City area.

New Ears Master List of Schools and Programs

This master list contains names, addresses, and phone numbers of over 400 audio-related programs from around the world. Programs are listed alphabetically by country, state, and name, resulting in a geographic order.

NRS Training School 9-13 Lawry Place
Macquarie ACT 2614 Australia *phone:* 06-251-6333

Australian Audio Education Centre 107 Union Street
North Sydney New South Wales 2060 Australia *phone:* 02-929-4944

Australian Film Television & Radio School Balaclava & Epping Roads, PO Box 126
North Ryde New South Wales 2113 Australia *phone:* 61-2-805-6611

Fat Boy Recording Studios 53 Victoria Road
Rozelle New South Wales 2039 Australia *phone:* 02-818-1716

JMC Academy of Audio Engineers 396 Elizabeth Street
Surrey Hills New South Wales 2010 Australia *phone:* 02-281-8899

Lismore College of Tafe 16-18 Carrington Street
Lismore New South Wales 2480 Australia *phone:* 066-21-9307

Music Futures PO Box 251
Vaucluse New South Wales 2030 Australia *phone:* 02-388-7222

Music Headquarters- Production Studios 16A Hirst Street
Jesmond New South Wales 2299 Australia *phone:* 049-50-1574

Professional Audio School 109 Sadlier Avenue
Ashcroft New South Wales 2168 Australia *phone:* 02-607-0494

R & R Recording & Rehearsal Unit 4/3 Anvil Road
Seven Hill New South Wales 2147 Australia *phone:* 02-624-4484

Riversound Recordings 339 Belmore Road
Riverwood New South Wales 2210 Australia *phone:* 02-534-3496

School of Audio Engineering- Sydney 68-72 Wentworth Avenue
Sydney New South Wales 2010 Australia *phone:* 61-2-21-13711

Sound Solution Shop 4/105 Union Street
Lismore New South Wales 2480 Australia *phone:* 066-21-7614

Atlantis Recording 1 Kingsway Recording
South Melbourne Queensland 3205 Australia *phone:* 03-629-8464

Audio Visual College 1-3 Gordon Street
Richmond Queensland 3121 Australia *phone:* 03-428-8812

Australian Academy of Music 122 Leichhardt Street
Spring Hill Queensland 4000 Australia *phone:* 07-831-0283

Australian Film Television & Radio School 40 George Street, 2nd Flr
Brisbane Queensland 4000 Australia *phone:* 61-7-221-3188

C'est Ca Audio-Visual Services Rear 242 Smith Street
Collingwood Queensland 3066 Australia *phone:* 03-419-1506

Moe Recording Services Lot 3 Willow Grove Road
Tanjil South Queensland 3825 Australia *phone:* 051-60-1485

Multi Media Productions 5/5 Shearson Cres
Mentone Queensland 3194 Australia *phone:* 03-583-7134

O'Brien Recordings 74 Bingalong Drive
Ashmore Queensland 4214 Australia *phone:* 075-39-3501

Pyramid Productions Audio Course Suite 235/20 Duncan Street
Fortitude Valley Queensland 4006 Australia *phone:* 07-852-1750

Queensland Conversvatorium of Music Music Technology Department, PO Box 28
North Quay Queensland 4002 Australia *phone:* 61-7-875-6256

School of Audio Engineering- Brisbane 22 Heussler Tce, Milton
Brisbane Queensland 4064 Australia *phone:* 61-7-369-8108

Australian Film Television & Radio School 20 Princes Street
Adelaide South Australia 5000 Australia *phone:* 61-8-232-4266

School of Audio Engineering- Adelaide 18-20 Deeds Road
Adelaide South Australia 5038 Australia *phone:* 08-376-0991

Australian Film Television & Radio School 47 Sandy Bay Road
Battery Point Tasmania 7004 Australia *phone:* 61-2-23-8703

Abbey Sound 23 Hartnet Drive
Seaford Victoria 3198 Australia *phone:* 03-786-4211

Australian Audio College 763 High Street
Preston Victoria 3042 Australia *phone:* 61-3-478-2153

Australian Film Television & Radio School 274 City Road
South Melbourne Victoria 3205 Australia *phone:* 61-3-690-7111

Royal Melbourne Institute of Technology 80 Victoria Street
Carlton Victoria 3053 Australia *phone:* 03-663-5611

School of Audio Engineering- Melbourne 80-86 Inkerman Street, St. Kilda
Melbourne Victoria 3182 Australia *phone:* 61-3-534-4403

Victoria Audio Education Centre 1-3 Gordon Street
Richmond Victoria 3121 Australia *phone:* 03-428-8812

Australian Film Television & Radio School 4/24 Thorogood Street
Victoria Park Western Australia 6100 Australia *phone:* 61-9-470-5330

School of Audio Engineering- Perth 42 Wickham Street
East Perth Western Australia 6000 Australia *phone:* 09-325-4533

SSL Promotions 316 Charles Street
North Perth Western Australia 6006 Australia *phone:* 09-227-6949

School of Audio Engineering- Vienna Leystr. 43
Vienna A-1200 Austria *phone:* 0222-3304133

Banff Centre for the Arts Box 1020
Banff Alberta T0L 0C0 Canada *phone:* 403-762-6651

Red Deer College School of Music
Red Deer Alberta T4N 5H5 Canada *phone:* 403-342-3216

Bullfrog Recording Studios 2475 Dunbar Street
Vancouver British Columbia V6R 3N2 Canada *phone:* 604-734-4617

Columbia Academy of Recording Arts 1295 W Broadway
Vancouver British Columbia V6H 3X8 Canada *phone:* 604-736-3316

Institute of Communication Arts 3246 Beta Avenue
Burnaby British Columbia V5G 4K4 Canada *phone:* 604-298-5400

Selkirk College 820 10th St
Nelson British Columbia V1L 3C7 Canada *phone:* 604-352-6601

Trebas Institute of Recording Arts 112 East 3rd Avenue
Vancouver British Columbia V5T 1C8 Canada *phone:* 604-872-2666

George Brown College PO Box 1015, Station B
Toronto Ontario Canada *phone:* 416-976-1212

Harris Institute for the Arts 296 King Street East
Toronto Ontario M5A 1K4 Canada *phone:* 416-367-0178

Ontario Institute of Audio Recording Technology 502 Newbold Street
London Ontario N6E 1K6 Canada *phone:* 519-686-5010

Recording Arts Program of Canada 28 Valrose Drive
Stoney Creek Ontario L8E 3T4 Canada *phone:* 416-662-2666

Trebas Institute of Audio Recording 410 Dundas Street, East
Toronto Ontario M5A 2A8 Canada *phone:* 416-966-3066

Trebas Institute of Recording Arts 440 Lauries Avenue West
Ottawa Ontario K1R 7X6 Canada *phone:* 613-782-2231

University of Waterloo Audio Research Group
Waterloo Ontario N2L 3G1 Canada *phone:* 519-885-1211

McGill University, Faculty of Music 555 Sherbrooke Street West
Montreal Quebec H3A 1E3 Canada *phone:* 514-398-4535

Recording Arts Program of Canada 34 Chemin des Ormes
Ste-Anne-des-Lacs Quebec J0R 1B0 Canada *phone:* 514-224-8363

Trebas Institute of Recording Arts 451 Saint Jean Street
Montreal Quebec H2Y 2R5 Canada *phone:* 514-845-4141

Aalborg University, Institute of Electronic Fredrik Bajers Vej 7
Aalborg DK-9220 Denmark *phone:* 45-8-15822

Danish Acoustical Institute Building 356, Akademives
Lyngby DK-2800 Denmark *phone:* 45-93-1244

Ecole Superierue de Realisation Audiovisuelle 135 Avenue Felix Faure
Paris 75015 France *phone:* 1-45-54-56-58

School of Audio Engineering- Paris 33, Rue le la Porte d'Aubervillers
Paris 75019 France *phone:* 43-1-330-4133

Universite Strasbourg, Directeurs du Son 22 rue Descartes
Strasbourg 67000 France *phone:*

Fachhochschule Düsseldorf, Fachberich Josef-Gockein-Strasse 9
40 Düsseldorf 4000 Germany *phone:* 02-11-4351-0

School of Audio Engineering- Berlin Seestr. 64
Berlin 65 1000 Germany *phone:* 030-456-5137

School of Audio Engineering- Frankfurt Homburger Landstrasse 182
Frankfurt 50 6000 Germany *phone:* 069-54-32-62

School of Audio Engineering- Munich Hofer Str. 3
Munich 83 8000 Germany *phone:* 089-67-51-67

Technical University Munchen Arcisstrasse 21, Electroacoustics
Munich D-8000 Germany *phone:* 49-89-2105-8541

Tonmeister der Hochschule fuer Music Allee 22
Detmold W-4930 Germany *phone:* 49-05231-740748

Hochschule der Kunste Berlin Fasanenstrasse 1-3
Berlin 12 D-1000 Germany *phone:* 030-31850

Film & Television Institute of India Law College Road
Pune Maharashtra 411 004 India *phone:* 331113

Entre Morale G. Feltrinelli Piazza Cantore 10
Milan 20123 Italy *phone:* 02-8323290

Kyushu Institute of Design 9-1 Shiobaru 4-chome
Minami-ku Fukuoka-shi 815 Japan *phone:* 092-553-4407

School of Audio Engineering- Kuala Lumpur Lot 5, Jalan 13/2 Petaling Jaya
Kuala Lumpur Selangor 46200 Malaysia *phone:* 03-756-7212

Koninklijk Conservatory Juliana van Stolberglaan 1
The Hague 2595 CA Netherlands *phone:* 31-70-381-42-51

School of Audio Engineering- Amsterdam Vondelstraat 13
Amsterdam 1054 GC Netherlands *phone:* 31-20-689- 4189

The Soundpeople 225 Barbadoes Street
Christchurch 1 New Zealand *phone:* 03-79-3393

School of Audio Engineering- Auckland 18 Heather Street
Parnell Auckland New Zealand *phone:* 9-373-4712

Frederic Chopin Academy of Music Okolnik 2
Warsaw Poland *phone:* 368-22-277241

Technical University of Gdansk Sound Engineering Dept., 80-952
ul. Majakowskiego Gdansk PL-80-952 Poland *phone:* 48-58 47-20-81

Royal Scottish Academy of Music & Drama School of Music, 100 Renfrew Street
Glasgow G2 3DB Scotland *phone:* 041-332 4101

Uniersity of Glasgow, Music Dept Gilmore Hill
Glasgow Scotland *phone:*

School of Audio Engineering- Glasgow 33 Coatbank Street
Coatbridge Glasgow ML5 3SP Scotland *phone:* 0236-436561

School of Audio Engineering- Singapore 122 Middle Road, #04-08 Midlink Plaza
Singapore 0716 Singapore *phone:* 65-3652523

Musikhögskolan i Piteå Box 210
Piteå 941 25 Sweden *phone:* 46-911-72627

BBC-Audio Engineering Unit BBC Wood Norton
Evesham WR11 4TF UK *phone:*

City of London Polytechnic Department of Music Technology
London EC3N 2EY UK *phone:*

City University, Music Department Northampton Square
London EC1V 0HB UK *phone:* 071-253-4399

London College of Furniture 41 Commercial Road, Instrument Tech
London UK *phone:* 01-247-1953

Media Production Facilities Bon Marche Bldg., Ferndale Road
London SW9 8EJ UK *phone:* 44-71-274-4000

Polytechnic of Central University School of Communication, 18-22 Riding House Street
London W1P 7PD UK *phone:* 071-486-5811

Polytechnic of North London Department of Electronic & Communication Engineering
London N7 8DB UK *phone:* 071-607-2789

Royal College of Music, Recording Studio Prince Consort Road
London SW7 2BS UK *phone:* 071-589-3643

Sandwell College Wednesbury
Sandwell WS12 0PE UK *phone:*

National Film & Television School Station Road
Beaconsfield Bucks HP9 1LG UK *phone:* 0494-678623

Open University Audio Visual Dept Walton Hall
Keynes Bucks MK7 6AA UK *phone:* 0908 74066

Campus AV- Professional Audio/Video Training 7 Miller Close, Offord Darcy
Huntingdon Cambridgeshire PE18 9SB UK *phone:* 0480 812201

School of Audio Engineering- London United House, North Road
London England N7 9DP UK *phone:* 44-71-6092653

Ravensbourne College of Design & Communication School of TV & Broadcasting
Bromley Kent BR1 3LE UK *phone:*

Salford College of Technology, Engineering Frederick Road
Salford Lancashire N6 6PU UK *phone:* 061-834-6633

Performing Arts & Technology School
Selhurst London UK *phone:*

Polytechnic of Central London School of Communications
18-22 Ridinghouse London W1P 7PD UK *phone:*

Thames Television Ltd. Teddington Lock
Teddinton Middlesex TW11 SNT UK *phone:* 01-977-3252

University of York, Dept. of Music and Electronics Heslington
York North Yorkshire Y01 5DD UK *phone:* 0904-432446

Keele University Electronic Music & Recording Studios
Keele Staffs. ST5 5BG UK *phone:* 0782-621111

University of Surrey Department of Music
Guilford Surrey GU2 5XH UK *phone:* 44-483-509317

Alabama State University 915 S. Jackson Street
Montgomery AL 36101 USA *phone:* 205-293-4346

Calhoun Community College PO Box 2216
Decatur AL 35609-2216 USA *phone:* 205-353-3102

Studio Four 1918 Wise Drive
Dothan AL 36303 USA *phone:* 205-794-9067

University of North Alabama Box 5040
Florence AL 35632-0001 USA *phone:* 205-760-4361

Western Washington University Department of Physics
Lexington AL 35648 USA *phone:* 205-247-3983

Academy of Radio Broadcasting 4914 E McDowell Rd #107
Phoenix AZ USA *phone:* 602-267-8001

Conservatory of Recording Arts & Sciences 1100 East Missouri, Ste 400
Phoenix AZ 85014 USA *phone:* 602-265-6383

Global Media Institute of Arts & Sciences 6325 N Invergordon
Paradise Valley AZ 85253 USA *phone:* 602-948-5883

Mesa Community College 1833 W Southern Avenue
Mesa AZ 85202 USA *phone:* 602-461-7573

Scottsdale Community College 9000 E Chaparral Rd
Scottsdale AZ 85250 USA *phone:* 602-423-6350

Southwest Institute of Recording Arts 4831 N. 11th Street, Ste. C
Phoenix AZ 85014 USA *phone:* 602-241-1019

University of Arizona School of Music
Tucson AZ 85721 USA *phone:* 602-621-1341

Academy of Radio Broadcasting 8907 Warner Ave #115
Huntington Beach CA USA *phone:* 714-842-0100

AFI-Apple Computer Center 2021 North Western Avenue
Los Angeles CA 90027 USA *phone:* 800-999-4AFI

American River College, Music Department 4700 College Oak Drive
Sacramento CA 95841 USA *phone:* 916-484-8420

Audio Institute of America PO Box 15427
San Francisco CA 94115 USA *phone:* 415-931-4160

Audio Recording Technology Institute 1325 Red Gun Street
Anaheim CA 92806 USA *phone:* 714-666-2784

Blue Bear School of Music Building D, Fort Mason
San Francisco CA 94123 USA *phone:* 415-673-3600

California Institute of Concert Sound Engineering 1733 S Douglas Rd, Ste F
Anaheim CA 92806 USA *phone:* 714-634-4131

California Institute of Technology MLAF, 102-31 Caltech
Pasadena CA 91125 USA *phone:* 818-356-4590

California Institute of the Arts, Music 24700 McBean Parkway
Valencia CA 91355 USA *phone:* 805-255-1050

California Polytechnic State University Music Department
San Luis Obispo CA 93407 USA *phone:* 805-756-2406

California Recording Arts Academy PO Box 1623
Sonoma CA 95476 USA *phone:* 707-996-4363

California Recording Institute 970 O'Brien Drive
Menlo Park CA 94025 USA *phone:* 415-324-0464

California State Polytech, Music Dept. 3801 West Temple Avenue
Ponoma CA 91768-4051 USA *phone:* 714 869-3548

California State University at Chico Music Department
Chico CA 95929-0805 USA *phone:* 916-898-5152

California State University at Dominquez Hills 1000 E Victoria Street, Music Department
Carson CA 90747 USA *phone:* 310-516-3543

California State University at Northridge Electical & Computer Engineering Department
Northridge CA 91324 USA *phone:* 213-885-1200

California State University- San Marcos Music Technology- VPA
San Marcos CA 92096 USA *phone:*

Citrus College 1000 W Foothill Blvd
Glendora CA 91740 USA *phone:* 818-914-8511

City College of San Francisco 50 Phelen Avenue
San Francisco CA 94112 USA *phone:* 415-239-3525

Cogswell Polytechnical College 10420 Bubb Road
Cupertino CA 95014 USA *phone:* 408-252-5550

College for Recording Arts 665 Harrison Street
San Francisco CA 94107 USA *phone:* 415-781-6306

Columbia College 925 North La Brea Avenue
Los Angeles CA 90038 USA *phone:* 213-851-0550

CSU Summer Arts Humbolt State University
Arcata CA 95521 USA *phone:* 707-826-5401

Finn Jorgensen Danvik 1201 Bel Air Drive
Santa Barbara CA 93105 USA *phone:* 805-682-2102

First Light Video Publishing 8536 Venice Blvd.
Los Angeles CA 90034 USA *phone:* 800-777-1576

Fullerton College, Music Department 321 East Chapman Avenue
Fullerton CA 92632 USA *phone:* 714-992-7296

Golden West College, Music Dept. 15744 Golden West Avenue
Huntington Beach CA 92647 USA *phone:* 714-895-8780

Institute for Audio/Video Engineering 1831 Hyperion Avenue
Hollywood CA 90027 USA *phone:* 213-666-2380

Long Beach City College 4901 E. Carson Street
Long Beach CA 90808 USA *phone:* 213-420-4309

Los Angeles City College 855 North Vermont Avenue
Los Angeles CA 91326 USA *phone:* 213-953-4521

Los Angeles Harbor College 111 Figueroa Place
Wilmington CA 90744 USA *phone:* 310-522-8200

Los Angeles Recording Workshop 12268-X Ventura Blvd.
Studio City CA 91604 USA *phone:* 818-763-7400

Los Medanos College 2700 East Leland Road
Pittsburgh CA 94565 USA *phone:* 415-439-2181

Loyola Marymount University Communication Arts Dept
Los Angeles CA 90045 USA *phone:* 310-338-3033

Media Workshops Foundation 734 North Bundy Drive
Los Angeles CA 90049 USA *phone:* 800-223-4561

Mendocino County Regional Occupation Program PO Box 226
Mendocino CA 95460 USA *phone:* 707-937-1200

Mills College- CCM 5000 MacArthur Blvd.
Oakland CA 94613 USA *phone:* 510-430-2191

Miracosta College Music Dept One Bernard Dr
Oceanside CA 92056 USA *phone:* 619-757-2121

Mus-i-col Recording Studios, RIA 9851 Prospect Avenue
Sanfee CA 92071 USA *phone:* 714-448-6000

Orange Coast College 2701 Fairview Rd, PO Box 5005
Costa Mesa CA 92628-5005 USA *phone:* 714-432-5692

Prairie Sun PO Box 7084
Cotati CA 94931 USA *phone:* 707-795-7011

Recording Insitute 14511 Delano Street
Van Nuys CA 91411 USA *phone:* 213-254-1756

S & S Sound Engineering Course PO Box 1156
Torrance CA 90505 USA *phone:* 213-375-0768

Sacramento City College 3835 Freeport Blvd
Sacramento CA 95822 USA *phone:* 916-558-2130

San Diego Recording Workshop PO Box 7632
San Diego CA 92167 USA *phone:* 619-571-5965

San Francisco State University 1600 Holloway Ave
San Francisco CA 94132 USA *phone:* 415-338-1372

San Francisco State University Extended 425 Market Street
San Francisco CA 94105 USA *phone:* 415-904-7720

San Jose State University, Electro-Acoustic One Washington square
San Jose CA 95192-0095 USA *phone:* 408-924-4773

Sonoma Sound PO Box 1623
Sonoma CA 95476 USA *phone:* 707-996-4363

Sonoma State University Department of Music
Rohnort Park CA 94928 USA *phone:* 707-664-2324

Sony Institute of Applied Video Technology 2021 N Western Avenue
Los Angeles CA 90029 USA *phone:* 213-462-1982

Sound Investment Enterprises PO Box 4139
Thousand Oaks CA 91359 USA *phone:* 805-499-0539

Sound Master Recording School 10747 Magnolia Blvd.
N. Hollywood CA 91601 USA *phone:* 213-650-8000

Stanford University, Music Department Center for Computer Research in Music
Stanford CA 94305-8180 USA *phone:* 415-723-3811

Trebas Institute of Recording Arts 6464 Sunset Boulevard
Los Angeles CA 90028 USA *phone:* 213-467-6800

UCLA Extension 10995 Le Conte Avenue, Room 437
Los Angeles CA 90099-2137 USA *phone:* 310-825-9064

University of California at Berkeley Center for New Music & Audio Technologies
Berkeley CA 94709 USA *phone:* 510-642-2678

University of California at Santa Cruz 1156 High Street
Santa Cruz CA 95064 USA *phone:* 408-459-2369

University of Southern California School of Music, MUS 409
Los Angeles CA 90089-0851 USA *phone:* 213-740-3217

Yamaha P.A.C.E. School 6600 Orangethorpe Avenue
Buena Park CA 90620 USA *phone:* 714-522-9474

Aims Community College 4801 W 20th St
Greeley CO 80634 USA *phone:* 303-8008

Colorado Institute of Art 200 E. 9th Street
Denver CO 80203 USA *phone:* 303-837-0825

Denver Public Schools, Career Education Center 2650 Eliot Street
Denver CO 80211 USA *phone:* 303-964-3075

Institute for Music, Health, and Education PO Box 1244
Boulder CO 80306 USA *phone:* 303-443-8484

University of Colorado at Denver, Music PO Box 173364, Campus Box 162
Denver CO 80217-3364 USA *phone:* 303-556-2727

University of Denver, Lamont School of Music 7111 Montview Blvd.
Denver CO 80220 USA *phone:* 303-871-6952

University of Northern Colorado School of Music
Greeley CO 80369 USA *phone:* 303-351-2678

Danbury Music Education Center 52A Federal Dr
Danbury CT 06810 USA *phone:* 203-743-5656

Media Arts Center 753 Capitol Avenue
Hartford CT 06106 USA *phone:* 203-951-8175

National Guitar Summer Workshop PO Box 222
Lakeside CT 06758 USA *phone:* 203-567-5829

Northwestern Connecticut Community College Park Place
Winsted CT 06098 USA *phone:* 203-738-6300

RBY Recording and Video 920 North Main Street
Southbury CT 06488 USA *phone:* 203-264-3666

Trod Nossel Recording, RIA 10 George Street, PO Box 57
Wallingford CT 06492 USA *phone:* 203-269-4465

University of Hartford, College of Engineering 200 Bloomfield Avenue
West Hartford CT 06117-0395 USA *phone:* 203-768-4792

University of New Haven 300 Orange Avenue
West Haven CT 06516 USA *phone:* 203-932-7101

Yale University School of Drama- Sound Design
New Haven CT 06520 USA *phone:* 203-432-1507

American University, Physics Department 4400 Massachusetts Avenue NW
Washington DC 20016-8085 USA *phone:* 202-885-2743

Art Institute of Ft. Lauderdale 1799 SE 17th Street
Ft. Lauderdale FL 33316 USA *phone:* 800-275-7603

Career Training Institutes, Music & Video 1205 Washington Avenue
Miami Beach FL 33139 USA *phone:* 305-531-3300

Florida Southern College Dept of Music
Lakeland FL 33801-5698 USA *phone:* 813-680-4217

Full Sail Center for the Recording Arts 3300 University Blvd., Suite 160
Winter Park FL 32792 USA *phone:* 407-679-6333

Josh Noland Music Studio 760 W. Sample Road
Pompano Beach FL 33064 USA *phone:* 305-943-9865

Miami Sunset Senior High School 13125 SW 72nd Street
Miami FL 33183 USA *phone:* 305-385-4255

Miami-Dade Community College, South Campus 11011 SW 104th Street, Room 8242
Miami FL 33176 USA *phone:* 305-237-2265

Recording Skills & Music Business Workshop 8795 SW 57th Street
Cooper City FL 33328-5930 USA *phone:* 305-434-1377

Sony Professional Audio Training Group 6500 N. Congress Avenue
Boca Raton FL 33487 USA *phone:* 407-998-6700

Unity Gain Recording Institute 2976-F Cleveland Avenue
Fort Meyers FL 33901 USA *phone:* 813-332-4246

University of Miami- Electrical Engineering College of Engineering
Coral Gables FL 33124-0640 USA *phone:* 305-284-3291

University of Miami- Music School of Music
Coral Gables FL 33124-7610 USA *phone:* 305-284-2245

University Of South Florida School of Mass Communications
Tampa FL 33620-7800 USA *phone:* 813-974-2851

Art Institute of Atlanta 3376 Peachtree Rd
Atlanta GA 30326 USA *phone:* 800-275-4242

Georgia Institute of Technology School of Electrical Engineering
Atlanta GA 30332-0250 USA *phone:* 404-894-3090

Georgia State University School of Music University Plaza
Atlanta GA 30303 USA *phone:* 404-651-3676

Music Business Institute 3376 Peachtree Road NE
Atlanta GA 30326 USA *phone:* 404-266-2662

Oracle Recording Studio PO Box 464188
Lawrenceville GA 30246 USA *phone:* 404-921-7941

Iowa State University 2019 Black Engineering Bldg
Ames IA 50011 USA *phone:* D.K. Holger

Artist Workshop Recording 228 East Maple Street
Kankakee IL 60901 USA *phone:*

Bradley University Dept of Music
Peoria IL 61625 USA *phone:* 309-677-2595

Columbia College of Chicago 600 S Michigan
Chicago IL 60605 USA *phone:* 312-663-1600

Creative Audio 705 Western Avenue
Urbana IL 61801 USA *phone:* 217-367-3530

DePaul University School of Music 804 West Belden Avenue
Chicago IL 60614 USA *phone:* 312-362-6844

Elmhurst College 190 Prospect
Elmhurst IL 60126 USA *phone:* 708-617-3515

Governors State University Division of Communications
University Park IL 60466-0975 USA *phone:* 708-534-5000

Greenville College 315 East College
Greenville IL 62246 USA *phone:* 800-345-4440

Millikin University 1184 West Main
Decatur IL 62522 USA *phone:* 217-424-6254

North Illinois University Physics Department
Dekalb IL 60115 USA *phone:* 815-753-6493

Private Studios 705 W Western Avenue
Urbana IL 61801 USA *phone:* 217-367-3530

University of Illinois Drama Dept- Krannert Center
Urbana IL 61801 USA *phone:* 217-333-2371

University of Illinois at Urbana-Champaign 1114 W Nevada, 2136 Music Bldg
Urbana IL 61801 USA *phone:* 217-244-1207

University of Illinois, CERL Sound Group 252 Engineering Research Laboratory
Urbana IL 61801-2977 USA *phone:* 217-333-0766

Western Illinois University School of Music
Macomb IL 61455 USA *phone:* 309-298-1464

Ball State University Music Engineering Technology Studios
Muncie IN 47306 USA *phone:* 317-285-5537

Butler University, Dept of Radio & TV 4600 Sunset Avenue
Indianapolis IN 46208 USA *phone:* 317-283-9501

Indiana State University Dept of Music
Terre Haute IN 47809 USA *phone:* 812-237-2771

Indiana University School of Music- Audio Department
Bloomington IN 47405 USA *phone:* 812-855-1087

Institute for Leasure Industry Studies PO Box 80063
Indianapolis IN 46280-0068 USA *phone:*

Purdue University Drama Dept- Theater Sound Design
West Lafayette IN 47907 USA *phone:*

Purdue University Calumet Department of Electrical Engineering Technology
Hammond IN 46323 USA *phone:* 219-989-2471

Synergetic Audio Concepts Rt. 1, Box 267
Norman IN 47264 USA *phone:* 812-995-8212

University of Evansville School of Music
Evansville IN 47722 USA *phone:* 812-479-2252

Valparaiso University Dept of Music
Valparaiso IN 46383 USA *phone:* 219-464-5362

Vincennes University 1002 North 1st Street
Vincennes IN 47591 USA *phone:* 812-885-4135

University of Iowa, School of Music 2057 Music Building
Iowa City Iowa 52242-1793 USA *phone:* 319-335-1664

Bethany College, Dept of Music 421 N First
Lindsburg KS 67456-1897 USA *phone:* 913-227-3311

Kansas State University Music Department
Manhattan KS 66502 USA *phone:* 913-532-5740

Sound & Video Contractor PO Box 12901
Overland Park KS 66212 USA *phone:* 913-888-4664

Sunset Productions Recording Workshop 117 West Eight Street
Hays KS 67601 USA *phone:* 913-625-9634

Eastern Kentucky University School of Music
Richmond KY 40475 USA *phone:* 606-622-3266

Loyola University Recording Studio College of Music, PO Box 82
New Orleans LA 70118 USA *phone:* 504-865-3037

Audio Workshop School 119 Fresh Pond Pkwy.
Cambridge MA 02138 USA *phone:* 617-547-3957

Berklee College of Music- MP & E 1140 Boylston Street
Boston MA 02215 USA *phone:* 617-266-1400

Berklee College of Music- Synthesis 1140 Boylston Street
Boston MA 02215 USA *phone:* 617-266-1400

Boston University Drama Dept- Technical Theater
Boston MA 02215 USA *phone:* 617-353-3390

Bruel & Kjaer Instruments, Inc. 185 Forest Street
Marlborough MA 01752 USA *phone:* 617-481-7000

Dean Junior College, Communication Arts 99 Main St
Franklin MA 02038 USA *phone:* 508-528-9100

MIT - Media Arts & Sciences 20 Ames Street, Rm E15-228
Cambridge MA 02139 USA *phone:* 617-253-5114

Northeast Broadcasting School 142 Berkeley St
Boston MA 02116 USA *phone:* 617-267-7910

Northeastern University, Music Department 360 Huntington Avenue
Boston MA 02115 USA *phone:* 617-437-2440

University of Massachusetts at Lowell One University Avenue, College of Fine Arts
Lowell MA 01854 USA *phone:* 508-934-3850

Omega Studio's School of Recording Arts & 5909 Fishers Lane
Rockville MD 20852 USA *phone:* 301-230-9100

Peabody Conservatory of the John Hopkins 1 East Mt. Vernon Pl.
Baltimore MD 21202 USA *phone:* 410-659-8136

Studio Techniques PO Box 714
Lanham MD 20706 USA *phone:* 301-552-2716

International Film & Television Workshops 2 Central Street
Rockport ME 04856 USA *phone:* 207-236-8581

New England School of Broadcasting One College Circle
Bangor ME 04401 USA *phone:* 207-947-6083

University of Maine at Augusta University Heights
Augusta ME 04330 USA *phone:* 207-621-3267

Central Michigan University 340 Moore Hall
Mt. Pleasant MI 48859 USA *phone:* 517-774-3851

Interlochen Center for the Arts PO Box 199
Interlochen MI 49643 USA *phone:* 616-276-9221

Lansing Community College 315 North Grand Avenue
Lansing MI 48901 USA *phone:* 517-483-1670

Michigan State University, Telecommunications 209 Comm Arts & Sciences Bldg
E Lansing MI 48824 USA *phone:* 517-355-8372

Michigan Technical University Department of Electical Engineering
Houghton MI 49931 USA *phone:* 906-487-2550

Oakland University / K & R Studios 28533 Greenfield
Southfield MI 48076 USA *phone:* 313-557-8276

Recording Institute of Detroit 14611 E Nine Mile Road
Eastpoint MI 48021 USA *phone:* 313-779-1380

Specs Howard School of Broadcasting 16900 W Eight Mile Rd, Ste 115
Southfield MI 48075-5273 USA *phone:* 313-569-0101

Western Michigan University School of Music, Sound Studios
Kalamazoo MI 49008 USA *phone:* 616-387-4720

Brown Institute 2225 E Lake Street
Minneapolis MN 55407 USA *phone:* 612-721-2481

Concordia College 901 s 8th St
Moorhead MN 56562 USA *phone:* 218-299-4202

Hennepin Technical College 9200 Flying Cloud Drive
Edin Prairie MN 55347 USA *phone:* 612-944-2222

Hutchinson Vocational Technical Institute 2 Century Avenue
Hutchinson MN 55350 USA *phone:* 612-587-3636

Moorhead State University Department of Music
Moorhead MN 56563 USA *phone:* 218-236-2101

Musicians Technical Training Center 304 North Washington Avenue
Minneapolis MN 55401 USA *phone:* 612-338-0175

Red Wing Technical Institute Highway 58
Red Wing MN 55066 USA *phone:* 612-388-8271

St. Paul Central High School 275 North Lexington
St. Paul MN 55104 USA *phone:* 612-645-9217

Northeast Missouri State University School of Music
Kirksville MO 63501 USA *phone:* 816-785-4439

School of the Ozarks Mass Media Department
Pt. Lookout MO 65726 USA *phone:* 471-334-6411

Smith/Lee Media School 7420 Manchester Road
St. Louis MO 63143 USA *phone:* 314-647-3900

University of Missouri at Kansas City 4949 Cherry, Performing Arts Center
Kansas City MO 64110 USA *phone:* 816-235-2964

Webster University 470 East Lockwood Avenue
St Louis MO 63119 USA *phone:* 314-968-6924

May Technical College 1306 Central Avenue
Billings MT 59102 USA *phone:* 406-259-7000

Appalachian State University School of Music
Boone NC 28608 USA *phone:* 704-262-3020

Barton College, Dept of Fine Arts College Station
Wilson NC 27893 USA *phone:* 919-399-6468

East Carolina University School of Music
Greenville NC 27858 USA *phone:* 919-757-6982

Elizabeth City State University Campus Box 809
Elizabeth City NC 27909 USA *phone:* 919-335-3377

University of North Carolina at Asheville UNCA Music Department
Asheville NC 28804-3299 USA *phone:* 704-258-6432

Weaver Education Center 300 S. Spring Street
Greensboro NC 27401 USA *phone:* 919-370-8282

Wingate College School of Music
Wingate NC 28174 USA *phone:* 704-233-8038

Minot State University Music Recording Studios
Minot ND 58701 USA *phone:* 701-857-3185

Kearny State College Dept of Music
Kearney NE 68849 USA *phone:* 308-234-8618

Northeast Community College 801 E Benjamin Avenue, PO Box 469
Norfolk NE 68701 USA *phone:* 402-644-0506

University of Nebraska-Lincoln Electrical Engineering, 209N WSEC
Lincoln NE 68588-0511 USA *phone:* 402-472-3771

Dartmouth College- Music Department 6147 Hopkins Center
Hanover NH 03755 USA *phone:* 603-646-3974

Brookdale Community College 765 Newman Springs Road
Lincroft NJ 07738 USA *phone:* 908-842-1900

Eastern Artists Recording Studio 36 Meadow Street
East Orange NJ 07017 USA *phone:* 201-673-5680

Egg Harbor Township High School English Creek & High School Drive
Egg Harbor Township NJ 08232 USA *phone:* 609-653-0100

Middlesex County College Woodbridge Avenue
Edison NJ 08817 USA *phone:* 908-906-2584

Sam Ash Music Institute 1077 Rt 1
Edison NJ 08837 USA *phone:* 908-549-0011

William Paterson College of New Jersey Music Department
Wayne NJ 07470 USA *phone:* 201-595-2315

New Mexico State University Music Dept., Box 30001-3F
Las Cruces NM 88003 USA *phone:* 505-646-2421

Quincy Street Sound 130 Quincy Street
Albuquerque NM 87108 USA *phone:* 505-265-5689

University of Nevada, Las Vegas 4504 S. Maryland Pkwy.
Las Vegas NV 89154 USA *phone:* 702-739-0819

Aspen Music Festival, Stanton Audio Institute 250 West 54th Street, 10th Floor East
New York NY 10019-5597 USA *phone:* 212-581-2196

Audio Recording Technology Institute 440 Wheeler Road
Hauppauge NY 11788 USA *phone:* 516-582-8999

Brooklyn College, CUNY Conservatory of Music
Brooklyn NY 11210 USA *phone:* 718-780-5582

Cayuga Community College 197 Franklin Street
Auburn NY 13021 USA *phone:* 315-255-1743

Center for Media Arts at Mercy College 226 West 26th Street
New York NY 10001 USA *phone:* 800-262-2297

College of St. Rose 432 Western Avenue
Albany NY 12203 USA *phone:* 518-454-5178

Dutchess Community College 53 Pendell Road
Poughkeepsie NY 12601-1595 USA *phone:* 914-471-4500

Eastman School of Music 26 Gibbs Street
Rochester NY 14604 USA *phone:* 716-274-1000

Finger Lakes Community College 4355 Lake Shore Drive
Canandaigua NY 14424-8395 USA *phone:* 716-394-3500

Finger Lakes Community College 4355 Lake Shore Dr
Canandaigua NY 14424 USA *phone:* 716-394-3500

Five Towns College 305 North Service Road
Dix Hills NY 11746-6055 USA *phone:* 516-424-7000

Herkimer County Community College 100 Reservoir Rd
Herkimer NY 13350-7995 USA *phone:* 315-866-0300

Hofstra University Department of Music
Hampstead NY 11550 USA *phone:* 516-463-5490

Institute of Audio Research 64 University Place
New York NY 10003 USA *phone:* 212-677-7580

Iona College-Seton School 1061 N. Broadway
Yonkers NY 10701 USA *phone:* 914-378-8024

Ithaca College, School of Communications 953 Danby Rd
Ithaca NY 14850 USA *phone:* 607-274-3242

Manhattan School of Music 120 Claremont Avenue
New York NY 10022 USA *phone:* 212-749-2802

Masterview Recording Studios 1621 Ithaca/Dryden Rd
Freeville NY 13068 USA *phone:* 607-844-4581

Media Center at Visual Studies Workshop 31 Prince Street
Rochester NY 14607 USA *phone:* 716-442-8676

Music Career Services
New York NY USA *phone:* 212-860-2082

New School 66 West 12 Street
Greenwich Village, NY 10011 USA *phone:* 800-544-1978

New York University, Department of Music 34 West 4th Street
New York NY 10003 USA *phone:* 212-998-5422

Rensselaer Polytechnic Institute iEAR Studios - DCC 135
Troy NY 12180-3590 USA *phone:* 518-276-4778

Rochester Institute of Technology Electrical Engineering Department
Rochester NY 14623 USA *phone:* 716-475-2174

Select Sound Studios 2315 Elmwood Avenue
Kenmore NY 14217 USA *phone:* 716-873-2712

Skidmore College Electronic Music Lab
Saratoga Springs NY 12866 USA *phone:* 518-584-5000

State University of New York at Oswego Lanigan Hall
Oswego NY 13126 USA *phone:* 315-341-2357

SUNY Fredonia Mason Hall, School of Music
Fredonia NY 14063 USA *phone:* 716-673-3221

SUNY Oneonta School of Music
Oneonta NY 13820 USA *phone:* 607-431-3415

SUNY Potsdam Crane School of Music
Potsdam NY 13676 USA *phone:* 315-267-2413

Syracuse University Newhouse School: Radio, TV, Film Department
Syracuse NY 13244 USA *phone:* 315-443-4004

The Kitchen, Haleakala, Inc. 512 West 19th Street
New York NY 10011 USA *phone:* 212-255-5793

Bowling Green State University College of Musical Arts
Bowling Green OH 43403 USA *phone:* 419-372-2181

Capital University Conservatory of Music 2199 E Main
Columbus OH 43209 USA *phone:* 614-236-6101

Cleveland Institute of Music 11021 East Blvd
Cleveland OH 44106 USA *phone:* 216-791-5000

International College of Broadcasting 6 S Smithville Rd
Dayton OH 45431-1833 USA *phone:* 513-258-8251

Lakeland Community College State Routes 306 and I-90
Mentor OH 44060 USA *phone:* 216-953-7000

Mansfield University Dept of Music
Mansfield OH 16933 USA *phone:* 717-662-4735

Oberlin Conservatory of Music Oberlin College
Oberlin OH 44074 USA *phone:* 216-775-8900

Ohio Northern University Freed Center
Ada OH 45810 USA *phone:* 419-772-1194

Ohio School of Broadcast Technology 5500 S Marginal Rd
Cleveland OH 44103-1098 USA *phone:* 216-431-5500

Ohio State University School of Music 1866 College Road
Columbus OH 43210 USA *phone:* 614-422-7899

Ohio University- Telecommunications 9 South College St
Athens OH 45701 USA *phone:* 614-593-4870

Recording Workshop 455 Massieville Road
Chillicothe OH 45601 USA *phone:* 800-848-9900

Southern Ohio College 1055 Laidlaw Avenue
Cincinnati OH 45237 USA *phone:* 512-242-3791

Southwestern Oklahoma State University School of Music
Weatherford OK 73096 USA *phone:* 405-774-3708

National Broadcasting School 2501 SW First Ave, Ste 101
Portland OR 97201-4797 USA *phone:*

Portland Community College, Music Department PO Box 19000
Portland OR 97219-0990 USA *phone:* 503-244-6111

Recording Associates 5821 SE Powell Blvd.
Portland OR 97206 USA *phone:* 503-777-4621

University of Oregon School of Music
Eugene OR 97403 USA *phone:* 503-346-3761

Alpha Wave Recording Studios 5042A West Chester Pike
Edgemont PA 19028 USA *phone:* 215-353-9535

Art Institute of Philadelphia 1622 Chestnut Street
Philadelphia PA 19103-5198 USA *phone:* 800-275-2474

Art Institute of Pittsburgh 526 Penn Ave
Pittsburgh PA 15222 USA *phone:* 800-275-4270

Clarion University of Pennsylvania Department of Music
Clarion PA 16254 USA *phone:* 814-226-2287

Duquesne University School of Music 600 Forbes Avenue
Pittsburgh PA 15282 USA *phone:* 412-434-5486

JTM Workshop of Recording Arts PO Box 606
Knox PA 16232 USA *phone:* 814-797-5883

Lebanon Valley College of Pennsylvania 101 North College Avenue
Annville PA 17003-0501 USA *phone:* 717-867-6181

Millersville University Dept of Music
Millersville PA 17551 USA *phone:* 717-872-3439

Music Factory Enterprises Ford & Washington Streets
Norristown PA 19401 USA *phone:* 215-277-9550

Pennsylvania State University 220 Special Services Building
University Park PA 16802 USA *phone:* 814-863-2911

Pennsylvania State University- Acoustics Graduate Program in Acoustics, Applied Science Bldg
University Park PA 16802 USA *phone:* 814-865-6364

Starr Recording Studio, RIA 210 Saint James Place
Philadelphia PA 19106 USA *phone:* 215-925-5265

University of Pennsylvania Moore School of Electrical Engineering
Philadelphia PA 19104 USA *phone:*

Widener University Music Department
Chester PA 19013 USA *phone:* 215-499-4338

Southeastern Media Institute, SCAC 1800 Gervais Street, Media Arts Center
Columbia SC 29201 USA *phone:* 803-734-8696

Strawberry Jamm Recording Studios 115 West College Street
Winnsboro SC 29180-1302 USA *phone:* 803-356-4540

South Dakota State University School of Music, Box 12212
Brookings SD 57007 USA *phone:* 605-688-4422

Belmont University School of Business
Nashville TN 37212 USA *phone:* 615-386-4504

Franklin Institute of Recording Sound Technology PO Box 1121
Franklin TN 37065 USA *phone:* 615-794-3660

Memphis State University CFA 232
Memphis TN 38152 USA *phone:* 901-678-2559

Middle Tennessee State University PO Box 21 MTSU
Murfreesboro TN 37132 USA *phone:* 615-898-2578

University of Tennessee at Martin Department of Music
Martin TN 38238 USA *phone:* 901-587-7405

University of Tennessee, Music Department 1741 Volunteer Blvd.
Knoxville TN 37996-2600 USA *phone:* 615-974-7552

Vanderbilt School of Music-Blair School of Music 2400 Blakemore Avenue
Nashville TN 37212 USA *phone:* 615-327-3680

Alvin Community College 3110 Mustang Road
Alvin TX 77511 USA *phone:* 713-331-6111

Art Institute of Dallas Two NorthPark, 8080 Park Lane
Dallas TX 75231 USA *phone:* 214-692-8086

Art Institute of Houston 1900 Yorktown
Houston TX 77056 USA *phone:* 800-275-4244

Audio Engineering Institute 2815 Swandale Drive
San Antonio TX 78230 USA *phone:* 512-344-3299

Austin Community College 11928 Stonehollow Drive
Austin TX 78758 USA *phone:* 512-832-4806

Baylor University- School of Music PO Box 97408
Waco TX 76798 USA *phone:* 817-755-1161

Cedar Valley College, Music Dept. 3030 North Dallas Avenue
Lancaster TX 75134 USA *phone:* 214-372-8128

Collin County Community College 2800 East Springcreek Pkwy
PLano TX 75074 USA *phone:* 214-881-5807

East Texas State Unviersity Music Department
Commerce TX 75428 USA *phone:* 214-886-5303

Houston CC, Commercial Music Dept. 901 Yorkchester
Houston TX 77079 USA *phone:* 713-468-6891

McLennan Community College 1400 College Drive
Waco TX 76708 USA *phone:* 817-750-3578

Midland College 3600 N. Garfield
Midland TX 79705 USA *phone:* 915-685-4648

Sam Houston College, RTF PO Box 2207
Huntsville TX 77341 USA *phone:* 409-294-1341

San Antonio College, Radio-TV-Film Dept. 1300 San Pedro Avenue
San Antonio TX 78284 USA *phone:* 512-733-2793

Skip Frazee Audio Engineering 3341 Towerwood, Suite 206
Dallas TX 75234 USA *phone:* 214-243-3735

Sound Arts Recording School 2036 Pasket
Houston TX 77092 USA *phone:* 713-688-8067

Sound Arts Recording Studio, RIA 2036 Pasket, Suite A
Houston TX 77092 USA *phone:* 713-688-8067

South Plains College 1401 S College Avenue
Levelland TX 79336 USA *phone:* 806-894-9611

Southern Methodist University Electronic Music Studios
Dallas TX 75275 USA *phone:* 214-692-2643

Southwest Texas State University Music Department, 601 University Drive
San Marcos TX 78666 USA *phone:* 512 245-2651

Texarkana College Recording Studios 2500 N. Robinson Road
Texarkana TX 75501 USA *phone:* 214-838-4541

University of North Texas, College of Music PO Box 13887
Denton TX 76203 USA *phone:* 817-565-4919

University of Texas at Arlington, Music Dept. UTA Box 19105
Arlington TX 76019 USA *phone:* 817-273-3471

University of Texas at Austin Music Dept- Sound Program
Austin TX 78712 USA *phone:*

University of Texas at San Antonio Division of Music
San Antonio TX 78285 USA *phone:* 512-691-4347

Brigham Young University Music Department
Provo UT 84602 USA *phone:* 801-379-3083

Alpha Studio- RIAA 2049 West Broad Street
Richmond VA 23220 USA *phone:* 804-358-3852

James Madison University Music Dept
Harrisonburg VA 22807 USA *phone:* 703-568-6197

Tidewater Community College 1700 College Cresent
Virginia Beach VA 23456 USA *phone:* 804-427-7294

University of Vermont Communication Studies
Burlington VT 05401 USA *phone:* 8021-656-3214

Art Institute of Seattle 2323 Elliott Avenue
Seattle WA 98121 USA *phone:* 800-345-0987

Barton Audio Recording School 4718 38th Ave NE
Seattle WA 98105 USA *phone:* 206-525-7372

Evergreen State College LIB 1326 TESC
Olympia WA 98505 USA *phone:* 206-866-6000

Horizon Recording Studio 1317 South 295 Place
Federal Way WA 98003 USA *phone:* 206-941-2018

National Broadcasting School 2615 Fourth Ave #100
Seattle WA 98121-1233 USA *phone:* 206-728-2346

Dave Kennedy Recording Studios 8006 West Appleton Avenue
Milwaukee WI 53218 USA *phone:* 414-527-3127

Trans America School 600 Williamson
Madison WI 53703 USA *phone:* 608-257-4600

University of Wisconsin- Oshkosh Music Department
Oshkosh WI 54901 USA *phone:* 414-424-4224

University of Wisconsin-Madison Interarts and Technology, School of Education
Madison WI 53706 USA *phone:* 608-262-2353

West Virginia Institute of Technology School of Music
Montgomery WV 25136 USA *phone:* 304-442-3229

Casper College, Music Department 125 College Drive
Casper WY 82601 USA *phone:* 307-268-2532

Taller de Arte de Sonoro 2 Da Calle de Campo Alegre, Qta. Petunia, Local 1, Chacao
Caracas 1060 Venezuela *phone:* 58-2-32-3149

New Ears Master List of Journals and Magazines

This master list contains names, addresses, FAX and phone numbers of over 100 audio-related journals and magazines from around the world. These periodicals are listed alphabetically by type and name.

AES Journal
3752 Thomas Point Road
Annapolis MD 21403

Audio Engineering
The Journal of the Audio Engineering Society

APRS News Digest
2 Windsor Square
Silver Street
Reading RG1 2TH United Kingdom

Audio Engineering
Association of Professional Recording Services News

ASA Journal
500 Sunnyside Blvd
Woodbury NY 11797-2999

Audio Engineering
The Journal of the Acoustical Society of America

Audio Media
Media House
Burrel Road, St. Ives
Chambridgeshire PE17 4LE United Kingdom

Audio Engineering
Europe's Leading Professional Audio Technology Magazine

Audio Vision & Prosound
PO Box 306
Cammeray NSW 2062 Australia

Audio Engineering
For Australia'a Corporate Communication

Audio Week
Warren Publishing
2115 Ward Court, NW
Washington DC 20037

Audio Engineering
The Authoritative News Service for the Audio Consumer Electronics Industry

Channels
Curraweena
Napolean Reef NSW 2795 Australia

Audio Engineering
The Professional Sound and Lighting Magazine

db, Sound Engineering Magazine
203 Commack Road, Ste 1010
Commack NY 11725

Audio Engineering
Serving the Recording, Broadcast and Sound Contracting Fields

DJ Times
Testa Communications
25 Willowdale Avenue
Port Washington NY 11050

Audio Engineering
The International Magazine for the Professional Mobile & Club DJ

E Q
2 Park Avenue, Ste 1820
New York City NY 10016

Audio Engineering
Project Recording and Sound Techniques

Home & Studio Recording
Music Maker Publications
21601 Devonshire St., Ste 212
Chatsworth CA 91311

Audio Engineering
The Magazine for the Recording Musician

Live MIX
6400 Hollis Street, Ste #12
Emeryville CA 94608

Audio Engineering
Professional Live Sound and Music Production

Live Sound
PO Box 12943
Overland Park KS 66282-2943

Audio Engineering
From the Publishers of Recording Engineer Producer

MIX
6400 Hollis Street, Ste #12
Emeryville CA 94608

Audio Engineering
Professional Recording, Sound, and Music Production

MIX Line
6400 Hollis Street, Ste #12
Emeryville CA 94608

Audio Engineering
The Faxest News in Pro Audio, a fax newsletter, Phone: 510-653-5142

NSCA & CEDIA Newsletters
10400 Roberts Road
Palos Hills IL 60465

Audio Engineering
Newsletters for the National Sound and Communications Association and the Custom Electronic Design and Installation Association

One To One
Ludgate House, 8th Flr.
245 Blackfriars Road
London SE1 9UR United Kingdom

Audio Engineering
The Mastering, Pressing, and Duplicating Magazine

Open Ear
6717 NE Marshall Road
Bainsbridge Island WA 98110

Audio Engineering
A publication dedicated to sound and music in health and education

Pro Sound News
2 Park Avenue, Ste 1820
New York City NY 10016

Audio Engineering
Professional Sound Production Industry Magazine

Pro Sound News International
2 Park Avenue, Ste 1820
New York City NY 10016

Audio Engineering
The European Version of Pro Sound News

Professional Sound
67 Mowat Avenue, #350
Toronto ON M6K 3E3 Canada

Audio Engineering
The Canadian Sound Magazine

Sound & Communications
Testa Communications
25 Willowdale Avenue
Port Washington NY 11050

Audio Engineering
The Magazine for Sound Contractors and Installors

Sound Engineer & Producer
7 Swallow Place
London WC1R 7AA United Kingdom

Audio Engineering
The Sound Business Magazine

Sound & Video Contractor
Intertec Publishing Corp
PO Box 12901
Overland Park KS 66282-2912

Audio Engineering
The Journal for Sound and Video Contractors

Studio
Ludgate House, 8th Flr.
245 Blackfriars Road
London SE1 9UR United Kingdom

Audio Engineering
The Studio Magazine Read by the Music Industry

Studio Sound
Ludgate House, 8th Flr
245 Blackfriars Road
London SE1 9UR United Kingdom

Audio Engineering
The British Sound Recording and Production Magazine

Tape/Disc Business
701 Westchester Avenue
White Plains NY 10604

Audio Engineering
Covering the US Duplication Industry

Audio Amateur
305 Union Street
PO Box 576
Peterborough NH 03458-0576

Audiophile/Consumer
The Journal of Audiophile Crafts

Audio Magazine
1633 Broadway
New York City NY 10019

Audiophile/Consumer
Consumer high fidelity audio magazine

Glass Audio
305 Union Street
PO Box 576
Peterborough NH 03458-0576

Audiophile/Consumer
For those interested in high quality audio reproduction using vacuum tube technology

Speaker Builder
305 Union Street
PO Box 576
Peterborough NH 03458-0576

Audiophile/Consumer
The Loudspeaker Journal

Stereo Review
Diamandis Communications
1633 Broadway
New York City NY 10019

Audiophile/Consumer
Consumer high fidelity audio magazine

Broadcast Engineering
PO Box 12901
Overland Park KS 66282-2901

Broadcasting
Technically-oriented broadcast magazine

Broadcast Engineering News
PO Box 204
Strawberry Hills NSW 2012 Australia

Broadcasting
The Australian broadcast engineering magazine

Encore
PO Box 1377
Darlinghurst NSW 2010 Australia

Broadcasting
Australian Film, Television, and Video Production

Feedback
1771 N Street NW
Washington DC 20036-2891

Broadcasting
Publication of the Broadcast Education Association, also publish the Journal of Broadcasting and Electronic Media

International Broadcasting
100 Avenue Road
London NW3 3TP United Kingdom

Broadcasting
The Equipment Industry Magazine

Line Up
27 Old Gloucester Street
London WC1N 3XX United Kingdom

Broadcasting
The Journal of the Institute of Broadcast Sound

NAB & BEA Journals
1771 N St NW
Washington DC 20036

Broadcasting
Broadcasting and Broadcast Education Journals

Radio World International
5827 Columbia Pike, Ste 310
Falls Church VA 22041

Broadcasting
Radio's Best Read Newspaper

SBE Newsletter
PO Box 20450
Indianapolis IN 46220

Broadcasting
The Society of Broadcast Engineers Newsletter

Jazz Educators Journal
13888 W 3rd Place
Golden CO 80401

Education
The Journal of the International Association of Jazz Educators

MEIEA Newsletter
SUNY Oneonta
School of Music
Oneonta NY 13820

Education
Music & Entertainment Industry Educators Association News

Music Educators Journal
1902 Association Drive
Reston VA 22091

Education
The Journal of the Music Educators National Conference

AfterTouch
1024 W Wilcox Avenue
Peoria IL 61604

Electronic Music
The Computer Music Coalition Magazine

CMA Newsletter- Array
Computer Music Association
PO Box 1634
San Francisco CA 94101-1634

Electronic Music
The Newsletter of the Computer Music Association

Computer Music Journal
MIT Press
55 Hayward Street
Cambridge MA 02142

Electronic Music
Technically-oriented journal from MIT

Electronic Musician
6400 Hollis Street, Ste #12
Emeryville CA 94608

Electronic Music
Electronic music, recording, computing, and construction

Daily Variety
1400 N Cahuenga Blvd
Hollywood CA 90028

Entertainment Trade
News from the entertainment industry

Hollywood Reporter
6751 W Sunset Blvd
Hollywood CA 90028

Entertainment Trade
News from the Hollywood entertainment industry

Variety
475 Park Avenue South
New York City NY 10016

Entertainment Trade

Afterimage
31 Prince Street
Rochester NY 14607

Film/Video
The journal of the Media Alliance

American Cinematographer
1782 N Orange Drive
Hollywood CA 90028

Film/Video
Covering the art of motion picture production

Film & Video Magazine
8455 Beverly Blvd, Ste 508
Los Angeles CA 90048

Film/Video
The Film and Video Production Magazine

In Motion Film & Video Magazine
421 Fourth Street
Anapolis MD 21403

Film/Video

InView
920 Broadway
New York City NY 11010

Film/Video
Professional Video From the Inside Out

Millimeter
826 Broadway
New York City NY 10003

Film/Video
The Magazine of Motion Picture and Television Production

On Production & Post-Production
17337 Ventura Blvd, Ste 226
Encino CA 91316

Film/Video
Media Production and Post-Production

POST
Testa Communications
25 Willowdale Avenue
Port Washington NY 11050

Film/Video
The International Magazine for Post-Production Professionals

Producer's Quarterly
Testa Communications
25 Willowdale Avenue
Port Washington NY 11050

Film/Video
For Creative Image and Sound Professionals

SMPTE Journal
595 West Hartsdale Avenue
White Plains NY 10607

Film/Video
The Journal of the Society of Motion Picture and Television Engineers

Video Systems
Intertec Publishing Corp
PO Box 12901
Overland Park KS 66282-2912

Film/Video
The official magazine of the International Television Association

Videography
PSN Publications
2 Park Avenue, Ste 1280
New York City NY 10016

Film/Video
The Magazine of Professional Video Production, Technology, and Applications

Videomaker
PO Box 4591
Chico CA 95927

Film/Video
Camcorders, Editing, Desktop Video, Audio and Video Production

LDI
135 Fifth Avenue
New York City NY 10010-7193

Lighting Production
Lighting Dimensions International: The Magazine for the Lighting Professional

ADAT News/First Reflection
3630 Holdrege Avenue
Los Angeles CA 90016

Manufacturer
Newsletters for Alesis products and the ADAT Network

Cross Talk
Brüel & Kjaer
185 Forest Street
Marlborough MA 07152

Manufacturer
Product news from Brüel and Kjaer

New Ways in Music Technology
3443 East Paris SE
PO Box 899
Grand Rapids MI 49512-0899

Manufacturer
Part of Yamaha's series of newspapers for music education

Production Sound Report
Audio Services Corp
10639 Riverside Drive
N Hollywood CA 91602

Manufacturer
Film and location sound product newsletter from Audio Services Corporation

Roland User's Group Magazine
7200 Dominion Circle
Los Angeles CA 90040

Manufacturer
Electronic music and pro audio product news from Roland

Tascam User Guide
New Media Publications
524 San Anselmo Avenue
San Anselmo CA 94640

Manufacturer
Audio, MIDI, and sound design product news from Tascam

Transmitter
6 Vista Drive
PO Box 987
Old Lyme CT 06371

Manufacturer
Product news from Sennheiser Electronic Corporation

AudioVisuelle Media
Finsensvej 80
Frederiksberg DK 2000 Denmark

Multimedia
News Magazine for Production, People, and Techniques in the Audio Visual Field, in Danish

AV Video Magazine
701 Westchester Avenue
White Plains NY 10604

Multimedia
Production and Presentation Technologies

New Media Magazine
901 Mariner's Island Blvd, Ste 365
San Mateo CA 94404

Multimedia
Multimedia Technologies for Desktop Computer Users

Bass Player Magazine
20085 Stevens Creek
Cupertino CA 95014

Music Performance
Personalities, performance, and techniques for bass playters

Drums & Drumming
20085 Stevens Creek
Cupertino CA 95014

Music Performance
Personalities, performance, and techniques for drummers

Guitar for the Practicing Musician
10 Midland Avenue, PO Box 1490
Port Chester NY 10573

Music Performance

Guitar Player Magazine
20085 Stevens Creek
Cupertino CA 95014

Music Performance
Personalities, performance, and techniques for guitars players

Keyboard Magazine
20085 Stevens Creek
Cupertino CA 95014

Music Performance
Personalities, performance, and techniques for keyboardists and electronic musicians

Leonardo Music Journal
2030 Addison Street, Ste 600
Berkeley CA 94704

Music Performance
The Journal of the International Society for the Arts, Sciences and Technology.

MIDI Guitarist
PO Box 75
Jacksonville OR 97530

Music Performance

Musician Magazine
1515 Broadway, 39th Floor
New York City NY 01930

Music Performance
Consumer music magazine from the publishers of Billboard

Percussive Notes
PO Box 25
Lawton KS 73502

Music Performance
The journal of the Percussive Arts Society

The Instrumentalist
200 Northfield Road
Northfield IL

Music Performance
Music performance and education magazine

ASCAP Playback
ASCAP
One Lincoln Plaza
New York City NY 10023

Music Trade
News from the American Society of Composers, Authors, and Publishers

Billboard Magazine
1515 Broadway
New York City NY 10036

Music Trade
The International Music and Entertainment Trade

BMI Music World
320 West 57th Street
New York City NY 10019

Music Trade
News from Broadcast Music International

Canadian Music Trade
67 Mowat Avenue, #350
Toronto ON M6K 3E3

Music Trade
Canadian music business magazine

Canadian Musician
67 Mowat Avenue, #350
Toronto ON M6K 3E3

Music Trade
Canadian music performance magazine

Cashbox
330 West 58th Street
New York City NY 10019

Music Trade

Ear Magazine
131 Varick Street, #905
New York NY 10013

Music Trade
Covering avant-garde music of all varieties, including classical, electronic, rock, and world music

Independent Music Guide
PO Box 3516
Carbondale IL 62902

Music Trade
Highlighting independent music production and recording

Inside the RIAA
1020 19th Street NW, Ste 200
Washington DC 20036

Music Trade
News from the Recording Industry Association of America

International Musician
1501 Broadway
New York City NY 10036

Music Trade
News from the American Federation of Musicians

Music Business
10 Farmfield Road
Great Tey, Colchester
Essex CO6 1AB United Kingdom

Music Trade
The Monthly Magazine for the Muscial Instrument and Allied Trade

Music Business International
2 Park Avenue, Ste 1820
New York City NY 10016

Music Trade
Covering the European music business industry

Music, Inc.
Maher Publications
180 W Park Avenue
Elmhurst IL 60126-3379

Music Trade
Covering the musical instrument retail industry

Music & Sound Retailer
Testa Communications
25 Willowdale Avenue
Port Washington NY 11050

Music Trade
The Newsmagazine for Musical Instrument and Sound Product Merchandisers

Musical Merchandise Review
100 Wells Avenue
PO Box 9103
Newton MA 02159

Music Trade
Covering musical instruments and related products

NARAS Journal / Grammy Magazine
303 N Glenoaks Blvd, Ste 140
Burbank CA 91502

Music Trade
Publications of the National Academy of Recording Arts and Sciences

PULSE
2500 Del Monte Street, Bldg. "C"
West Sacramento CA 95691

Music Trade
Tower Record's magazine of recorded music

Rolling Stone
1290 Avenue of the Americas
New York City NY 10104-0298

Music Trade
Music personalities, politics, and promotion

Fast Forward
2219 W Olive Ave, Ste 264
Burbank CA 91506

Music Trades
Reviews demo recordings

Option
Sonic Options Network, Inc
2345 Westwood Blvd #2
Los Angeles CA 90064

Music Trades
Reviews alternate music from independent labels

Dramatics
3368 Central Parkway
Cincinnati OH 45225

Theater Production
The magazine of the Theatre Education Association and the International Thespian Society

TCI
135 Fifth Avenue
New York City NY 10010-7193

Theater Production
Theater Crafts International: The Business of Entertainment Technology and Design

New Ears Master List of Professional Trade Associations

This master list contains names, addresses, FAX and phone numbers of 75 audio-related professional trade associations from around the world. These organizations are listed alphabetically by name.

Academy of Motion Picture Arts and Sciences
8949 Wilshire Blvd
Beverly Hills CA 90211
Phone: 310-278-8990
Fax:

Academy of Television Arts and Sciences
3500 West Olive Ave, Ste 700
Burbank CA 90028
Phone: 818-953-7575
Fax:

Acoustical Society of America
500 Sunnyside Blvd
Woodbury NY 11797
Phone: 516-576-2357
Fax: 516-349-7669

Affiliated Independent Record Companies
PO Box 241648
Los Angeles CA 90024
Phone: 213-208-2104
Fax:

American Association for Music Therapy
66 Morris Avenue
Springfield NJ 07080
Phone: 201-379-1100
Fax:

American Council for the Arts
1285 Avenue of the Americas
New York NY 10019
Phone: 212-245-4510
Fax:

American Electronics Association
5201 Great America Parkway
Santa Clara CA 95054
Phone: 408-987-4200
Fax:

American Federation of Musicians
1501 Broadway, Ste 600
New York NY 10036
Phone: 212-869-1330
Fax: 212-764-6134

American Film Institute
AFI-Apple Computer Center
Los Angeles CA 90027
Phone: 213-856-7690
Fax: 800-999-4AFI

American Loudspeaker Manufacturers Association
3413 N Kennicott Avenue, Suite B
Arlington Hgts IL 60004
Phone: 312-577-7200
Fax:

American Society of Composers, Authors, and Publishers- Hollywood
6430 Sunset Blvd, Ste 200
Hollywood CA 90028
Phone: 213-466-7681
Fax: 213-466-6677

American Society of Composers, Authors, and Publishers- Nashville
66 Music Square West
Nashville TN 37203
Phone: 615-320-1211
Fax: 615-327-0314

American Society of Composers, Authors, and Publishers- New York
One Lincoln Plaza — *Phone:* 212-595-3050
New York NY 10023 — *Fax:* 212-274-9064

American Women in Radio and Television
1101 Connecticut Avenue NW — *Phone:* 202-429-5102
Washington DC 20036-4303 — *Fax:*

Association for Educational Communicatons and Technology
1025 Vermont Ave NW, Suite 820 — *Phone:* 202-347-7834
Washington DC 20005 — *Fax:* 202-347-7839

Association for International Practical Training
10 Corporate Center — *Phone:* 301-997-2200
Columbia MD 21044-3510 — *Fax:*

Association of Professional Recording Services
2 Windsor Square — *Phone:* 0734-756218
Reading RG1 2TH United Kingdom — *Fax:* 0734-756216

Audio Engineering Society- Europe
Zevenbunderslaan 142/9 — *Phone:* +32-2-345.7971
Brussels Belgium — *Fax:* +32-2-345.3419

Audio Engineering Society- UK
Lent Rise Road — *Phone:*
Slough SL1 7NY United Kingdom — *Fax:*

Audio Engineering Society- USA
60 East 42nd Street, Rm 2520 — *Phone:* 212-661-8528
New York NY 10165 — *Fax:* 212-682-0477

British Record Producers Guild
2 Windsor Square, Silver Street — *Phone:* 0734-756218
Reading RG1 2TH — *Fax:* 0734-756216

Broadcast Education Association
1771 N Street NW — *Phone:* 202-429-5355
Washington DC 20036-2891 — *Fax:*

Broadcast Music International- Los Angeles
8730 Sunset Blvd — *Phone:* 213-659-9109
Los Angeles CA 90069 — *Fax:*

Broadcast Music International- Nashville
10 Music Square East — *Phone:* 615-259-3625
Nashville TN 37203 — *Fax:*

Broadcast Music International- New York
320 West 57th Street
New York NY 10019
Phone: 212-586-2000
Fax:

Canadian Academy of Recording Arts & Sciences
124 Merton Street, 3rd Floor
Toronto ON M4S 2Z2 Canada
Phone: 416-485-3135
Fax:

Canadian Council on the Arts
PO Box 1047
Ottowa ON K1P 5V8
Phone: 613-598-4365
Fax:

Canadian Independent Record Production Association
144 Front Street West, Ste 202
Toronto ON M5J 2L7 Canada
Phone: 416-593-1665
Fax: 416-593-7563

Canadian Recording Industry Association
1255 Young Street, Ste 300
Toronto ON M4T 1W6 Canada
Phone: 416-967-7272
Fax: 416-967-9415

Career College Association
750 First St NE Ste 900
Washington DC 20002
Phone: 202-336-6700
Fax:

College Music Society
144 15th Street
Boulder CO 80302
Phone: 303-449-1611
Fax:

Computer Music Association
2040 Polk, PO Box 1634
San Francisco CA 94101-1634
Phone: 817-566-2235
Fax:

Computer Music Coalition
1024 W Willcox Avenue
Peoria IL 61604
Phone: 309-685-4843
Fax: 309-685-4878

Consumer Electronics Group
2001 I Street NW
Washington DC 20006
Phone: 202-457-8700
Fax:

Custom Electronic Design & Installation Association
10400 Roberts Road
Palos Hills IL 60465
Phone: 800-CEDIA30
Fax: 708-598-4888

Electronic Industries Association
2001 I Street NW
Washington DC 20006
Phone: 202-457-4900
Fax:

European Broadcast Union
Avenue Albert Lancaster 32 — *Phone:* 37-55-990
Brussels B-1180 Belgium — *Fax:*

IEEE Acoustics, Speech, & Signal Processing Society
455 Hoes Lane — *Phone:*
Piscataway NJ 08855 — *Fax:*

Independent Media Producers Council
3150 Spring Street — *Phone:* 703-273-7200
Fairfax VA 22031 — *Fax:*

Independent Music Network
PO Box 3516 — *Phone:* 618-549-8373
Carbondale IL 62902 — *Fax:*

Institute of Electrical & Electronic Engineers
345 East 47th Street — *Phone:* 212-705-7900
New York NY 10017 — *Fax:*

Internation Society for the Arts, Sciences and Technology
2030 Addison Street, Ste 600 — *Phone:*
Berkeley CA 94704 — *Fax:* 510-841-6311

International Alliance of Theatrical & Stage Employees
1515 Broadway, Ste 601 — *Phone:* 212-730-1770
New York NY 10036 — *Fax:*

International MIDI Association
5316 W 57th Street — *Phone:* 310-649-MIDI
Los Angeles CA 90056 — *Fax:* 310-215-3380

International Tape/Disc Association
505 Eighth Avenue — *Phone:* 212-643-0620
New York NY 10018 — *Fax:*

MIDI Manufacturers Association
5316 W 57th Street — *Phone:* 310-649-MIDI
Los Angeles CA 90056 — *Fax:* 310-215-3380

Music Educators National Conference
1902 Association Drive — *Phone:* 703-860-4000
Reston VA 22091 — *Fax:*

Music & Entertainment Industry Educators Association
Janet Nepke — *Phone:* 607-431-3425
Oneonta NY — *Fax:*

Music Industry Educators Association
1435 Bleury Street, Ste 301 — *Phone:*
Montreal HG3A 2H7 Canada — *Fax:*

Musical Instrument Technicians Association
3637 E 7800 South — *Phone:*
Salt Lake City UT 84121 — *Fax:*

National Academy of Recording Arts and Sciences
303 N Glenoaks Blvd, Suite 140 — *Phone:* 213-849-1313
Burbank CA 91502-1178 — *Fax:* 213-849-2529

National Affiliated Music Business Institutions
Dept of Music — *Phone:* 812-237-2771
Terre Haute IN 47809 — *Fax:*

National Association of Broadcasters
1771 N Street NW — *Phone:* 202-429-5300
Washington DC 20036 — *Fax:* 202-429-5343

National Association of College Broadcasters
Brown University- Box 1955 — *Phone:* 401-863-2225
Providence RI 02912 — *Fax:*

National Association of Independent Record Distributors and Manufacturers
6935 Airport Higway Lane — *Phone:* 609-655-6636
Pennsauken NJ 08109 — *Fax:*

National Association of Music Merchandisers
5140 Avenida Encinas — *Phone:* 619-438-8001
Carlsbad CA 92008 — *Fax:* 619-438-7327

National Association of Record Merchandisers
3 Eves Drive, Ste 307 — *Phone:* 609-569-2221
Marlton NJ 08053 — *Fax:*

National Association of Schools of Music
11250 Roger Bacon Drive — *Phone:* 703-437-0700
Reston VA 22090 — *Fax:*

National Cable Television Association
1724 Massachusetts Avenue NW — *Phone:*
Washington DC 20036 — *Fax:*

National Endowment for the Arts, Music Program
1100 Pennsylvania Ave NW, #702 — *Phone:* 202-682-5445
Washington DC — *Fax:* 202-682-5612

National Federation of Local Cable Programmers
906 Pennsylvania Avenue SW — *Phone:*
Washington DC 20003 — *Fax:*

National Music Publishers Association
205 E 42nd Street — *Phone:* 212-370-5330
New York NY 10017 — *Fax:*

National Sound and Communications Association
10400 Roberts Road — *Phone:* 800-466-NSCA
Palos Hills IL 60465 — *Fax:* 708-598-4888

Recording Industry Association of America
1020 19th Street NW, Ste 200 — *Phone:* 202-775-0101
Washington DC 20036 — *Fax:* 202-775-7253

Society of Broadcast Engineers
PO Box 20450 — *Phone:* 317-253-0122
Indianapolis IN 46220 — *Fax:*

Society of Cable Television Engineers
669 Exton Commons — *Phone:* 215-363-6888
Exton PA 19341 — *Fax:* 215-363-5898

Society of Composers, Authors, and Music Publishers of Canada
41 Valleybrook Drive — *Phone:* 416-445-8700
Don Mills ON M3B 2S6 — *Fax:* 416-445-7108

Society of European Stage Authors and Composers
156 W 56th Street — *Phone:* 212-586-3450
New York NY 10019 — *Fax:*

Society of Motion Picture and Television Engineers
595 W Hartsdale Avenue — *Phone:* 914-761-1100
White Plains NY 10607 — *Fax:* 914-761-3115

Society of Professional Audio Recording Services
4300 10th Avenue North — *Phone:* 407-641-6648
Lake Worth FL 33461 — *Fax:* 407-642-8263

Sound and Communication Industries Federation
4-B High Street — *Phone:* 010-44-0-628-66
Slough SL1 7JH United Kingdom — *Fax:* 010-44-0-628-65

Volunteer Lawyers for the Arts
1285 Avenue of the Americas — *Phone:* 212-977-9270
New York NY 10019 — *Fax:*

Women in Music

PO Box 441, Radio City Station — *Phone:* 212-627-1240
New York NY 10101 — *Fax:*

Women's Independent Label Distribution Network

5505 Delta River Drive — *Phone:* 517-323-4325
Lansing MI 48906 — *Fax:*

Women's Technet

PO Box 966 — *Phone:* 708-485-5161
Ukiah CA 95482 — *Fax:* 708-485-5373

New Ears Additional Resources

This section lists audio research centers, special organizations, and additional information not included elsewhere in New Ears.

Audio/Media Research Centers

The following are involved in research, development, or experimental application of audio, music, and/or media technology. Many of these organizations offer opportunities for graduate study or advanced research and development. Centers are listed alphabetically by their institutional affiliation.

American Film Institute, AFI-Apple Computer Center, Los Angeles, CA 90027, phone: 213-856-7690

California Institute of the Arts, Center for Experiments in Art, Information and Technology, 24700 McBean Parkway, Valencia, CA 91355, phone: 805-255-1050

Case Western Reserve University,Center for Music and Technology, Department of Music, Haydn Hall Cleveland, OH 44106

Rensselaer Polytechnic Institute, Integrated Electronic Arts Studios, DCC 135, Troy, NY 12180-3590, phone: 518-276-4778

Institute for Music, Health, and Education, PO Box 4179, Boulder, CO 80306, phone: 303-443-8484

IRCAM, 31 rue Saint-Merri, 75004 Paris, France, phone: 42-77-12-33, ext. 48-21

Massachusetts Institute of Technology, Media Lab, Cambridge, MA 02139, phone: 617-253-0300

Museum of Television and Radio, 25 West 52nd Street, New York, NY, phone: 212-621-6800

North Carolina State University, Center for Sound And Vibration, Campus Box 7910, Raleigh, NC 27695

Stanford University, Center for Computer Research in Music and Acoustics, Department of Music, Stanford, CA 94305, phone: 415-723-3811

Syracuse University, Institute for Sensory Research, Syracuse, NY 13244-5290, phone: 315-443-9749

University of California, Berkley, Center for New Music and Audio Technologies, 1750 Arch Street, Berkley, CA 94709, phone: 510-642-2678

University of California, San Diego, Computer Audio Research Laboratory, Center for Music and Related Research, La Jolla, CA 92093

University of Hartford, Acoustics & Vibrations Laboratory, College of Engineering, West Hartford, CT 06117, phone: 203-768-4792

University of North Texas, Center for Experimental Music and Intermedia, School of Music, Denton, TX 76205

University of Waterloo, Audio Research Group, Waterloo, Ontario, N2L 3G1 Canada, phone: 519-885-1211

Special Organizations

Here are a few items not included elsewhere in New Ears, yet deserving of recognition.

Electronic Bulletin Boards/Networks

America On-Line, Joe Zobkiw, 8619 Westwood Drive, Vienna, VA 22182, phone: 800-227-6364

Compuserve, 5000 Arlington Centre Blvd, Columbus, OH 43220, phone: 800-848-8990

Mac Users at Berklee BBS, Richard Boulanger, Berklee College of Music, 150 Massachusetts Avenue, Boston, MA 02115, phone: 617-266-1400

Music Network USA, Los Angeles, CA, phone: 310-312-8753

Performing Arts Network (PAN), PO Box 162, Skippack, PA 19474, phone: 215-584-0300

Sound-Net, 164 Sunnyside Ave, Ste 100, Toronto, Ontario, M6R 2P6 Canada, phone: 416-530-4423

Taxi, 5535 Canoga Ave, Ste 308, Woodland Hills, CA 91367, phone: 800-458-2111

Women's Technet, Vanessa Else, Silk Media, PO Box 966, Ukiah, CA 95482, phone: 707-485-5161.

Hearing Associations

HEAR: Hearing Education and Awareness for Rockers, PO Box 460847, San Francisco, CA 94146, phone: 415-773-9590. HEAR's mission is prevent hearing loss by promoting awareness of hearing issues and educating the general public about the nature of sound.

House Ear Institute, 2100 West Third Street, Los Angeles, CA 90057, phone: 213-483-4431. HEI is a private, non-profit corporation dedicated to uncovering the mysteries of the ear through hearing research studying the cause of impairment, diagnosis, and treatment. They train ear specialists and professionals from allied disciplines in diagnosis, treatment, and rehabilitative techniques.

National Information Center on Deafness, Gallaudet University, 800 Florida Avenue NE, Washington, DC 20002-3695, phone: 202-651-5051. NICD serves as a central source for up-to-date, objective information on topics dealing with hearing loss and deafness. They collect and disseminate information about all aspects of hearing loss and services offered to the deaf and hard of hearing people across the United States.

Public Access Studios/Alliances

Film/Video Arts, 817 Broadway, New York, NY 10003-4797, phone: 212-673-9361

Harvestworks/Studio Pass, 596 Broadway, New York, NY 10012, phone: 212-431-1130

Media Alliance, c/o Thirteen/WNET, 356 West 58th Street, New York, NY 10019, phone: 212-560-10019

NAMAC, 1212 Broadway, #816, Oakland, CA 94612, phone: 510-451-2717.

Visual Studies Workshop, 31 Prince Street, Rochester, NY 14607, phone: 716-442-8676

Check with your local cable system for public access groups in your area. Some offer a variety of training as well as practical experience.

Multimedia Programs: A Glimpse Into the Future

While New Ears has focused on programs related to audio, it is important to remember that the traditional borders between various forms of art and electronic media are beginning to vanish as more creative and technical production moves to common digital platforms. Desktop publishing and digital audio/video recording and editing are all merging on digital media workstations (DMWs ?). Though it may represent the state-of-the-art today, it will be the standard early tomorrow.

Many audio programs, recognizing this evolution, are incorporating multimedia classes into their curriculums. Several audio/media research centers are active with research and development in this area. The following are a sample of what the future holds.

American Film Institute Advanced Technology Programs, AFI-Apple Computer Center, 2021 North Western Avenue, Los Angeles, CA 90027, phone: 800-999-4AFI. The American Film Institute is a national trust dedicated to preserving the heritage of film and television, to identifying, developing, and training creative individuals, and to presenting the moving image as an art form. With the inauguration of the AFI-Apple Computer Center for Film and Videomakers in 1991, the AFI made a commitment to incorporate desktop digital tools into training film, video, and other media artist. The result has been a growing community of professionals from within and outside the traditional media arts. The AFI Advanced Technology Programs bring together under one umbrella all activities which explore digital technology in media. Seminars and workshops are offered in a variety of topics, including: pre-production, computer imaging, desktop video and film, computer animation, and interactive technology.

Massachusetts Institute of Technology Media Arts and Sciences Program, Room E15-224, MIT, Cambridge, MA 02139-4307, phone: 617-253-5114. At MIT, the phrase Media Arts and Sciences signifies the study, invention, and creative use of new information technologies in the service of human expression, education, and communication. The field is rooted in modern communication, computer, and human sciences, and its academic programs are linked with a wide range of research within MIT's Media Lab. Degree programs are offered at three levels: an undergraduate major, a master of science program, and a doctoral program. The programs' activities are divided into three sections: the Information and Entertainment Section, the Learning and Common Sense Section, and the Perceptual Computing Section. Within these, the academic and research agendas of individual faculty and researchers reflect various aspects of the science, technology, and aesthetics of human communication and the human/machine interface. Of interest to musicians and audio practitioners, is research areas in musical instrument invention and interactive performance. The Media Lab represents the cutting edge.

Integrated Electronic Arts at Rensselaer, iEAR Studios, DCC 135, Rensselaer Polytechnic Institute, Troy, NY 12180-3590, phone: 518-276-4778. The Department of Arts at Rensselaer offers a Master of Fine Arts degree in Electronic Arts. The program is conceived as an electronic arts program in an integrated artistic and technological environment, offering the opportunity to gain knowledge with computer music, video art, computer imaging, animation, and performance. The curriculum stresses creative studio-based work and emphasizes the unique problems presented by performance and public presentation of these media. The technological environment at Rensselaer offers ample opportunities for individual research projects, and the degree culminates in a large scale thesis project in the final year of study.

Interarts and Technology at the University of Wisconsin-Madison, School of Education, University of Wisconsin-Madison, 1050 University Avenue, Madison, WI 53706, phone: 608-262-2353. Interarts and Technology is a unique program integrating computer technologies with the visual, sound, and dance arts. The four-year program prepares students for graduate studies and expanding career opportunities in the arts, education, communications, and the electronic media industries. The curriculum is aimed at students with backgrounds in one or more of the arts who are interesting in applying computer technologies to the study of images, sound, and dance. Each core course in the program examines the relationship between the various art media.

A Parting Profile: For the Love of Audio

by Keith C. Seppanen

One day, at the age of sixteen, I was contemplating my future while taking a break from my math homework. A career in music ... that was an easy decision. My limited musical experience let me believe there were only two options; performer or band director. As a tuba player I did not see the bright lights of performance so that left me to pursue a career as a band director. Breaks over ... time to put on another stack of records and return to my math. Unaware of the music that was flowing into my head I carefully calculated each problem, until the LP Jonathan Livingston Seagull, by Neil Diamond, began to play. I decided to take another break and leaf through the color booklet that accompanied the two record set. I came to a page that had a picture of the sound engineer, Armin Stiener. That got me thinking how much I enjoyed technology and music ... what a great combination. My decision was made, audio engineering was the career for me.

Before graduating high school I did some research on careers in audio engineering. Unfortunately references like New Ears were not available and information was limited. I did find that I would need a strong musical background with a good balance of electronics. Not knowing where I could get this kind of education I entered college as a dual major; music and electrical engineering without practical experience in audio. Two years later and continuing my research I found that there were several colleges and universities that offered some type of degree in audio engineering. After painstaking review of each program I selected a new school and transferred the following year.

My education was valuable and well worth the time and money spent. I had excellent professors who brought their practical experience, knowledge, and love for the industry to the classroom. There was the late night multi-track recording sessions, sound reinforcement and live recording of university performances that offered a practical hands on education. There were many visiting industry professionals who seemed eager to share what was happening in the real world that helped to reinforce what I had learned in the classroom. One thing that haunted me from each of these visiting professionals was their consistent message; there are not an abundant number of jobs, the pay is low, the long hours, and forget having a personal relationship. Am I ready for graduation?

I did graduate and was able to step out into the audio industry. After many interviews and many rejections I was able to find a job as an assistant engineer in a recording studio. It was at this time that I learned that my prized bachelor's degree was only the foundation of my education. I had to continue to learn things like how to fill out work orders, making food runs, soldering cables, and the importance of team work. I eventually worked my way to chief engineer in the studio. After five and one-half years I decided to try my hand as an independent recording engineer. The education continued: I had to find my own clients, negotiate the rate of pay, collect the money, pay the quarterly taxes, learn the importance of team work even more, and get a great sound. When I decided to manage a recording studio I had to incorporate all that I had learned previously plus scheduling of clients and personnel. Currently I am teaching university students the art of recording. Hopefully, I am passing onto each student the knowledge I have received through my education along with my practical learning so that they will find their transition into the real world easier than mine. My own education has not yet stopped.

I feel that I have been successful and blessed in an industry that I love. Perhaps the love of the audio industry is why I have been able succeed. If you have a passion for your career the negatives you meet along the way don't seem so dark but are only hurdles in the pursuit of your dreams. There are jobs out there if you are willing to look, be flexible, and persevere. You can make enough money to live a comfortable life and the hours put in are only a commitment to the passion. You can have a personal relationship if you can find a partner who understands your love affair with audio.

Chase your dreams because dreams can become a reality.

Keith Seppanen is an award-winning music recording engineer and educator. Artists he has recorded include: Anita Baker, Chicago, the Crusaders, Debarge, Whitney Houston, Jermaine Jackson, Eddie Murphy, Lee Ritenour, David Sanborn, James Taylor, the Temptations, and Stanley Turrentine.